THE FIGHT OVER FOOD

THE FIGHT OVER
FOOD

Producers, Consumers, and Activists
Challenge the Global Food System

**Edited by
Wynne Wright and Gerad Middendorf**

The Pennsylvania State University Press
University Park, Pennsylvania

Library of Congress Cataloging-in-Publication Data

The fight over food : producers, consumers, and activists challenge the global food system /
[edited by] Wynne Wright and Gerad Middendorf.
p. cm. — (Rural studies series)
Includes bibliographical references and index.
ISBN 978-0-271-03274-0 (cloth : alk. paper)
ISBN 978-0-271-03275-7 (pbk. : alk. paper)
1. Food supply—Social aspects.
2. Agriculture–Social aspects.
I. Wright, Wynne.
II. Middendorf, Gerad.
III. Series: Rural studies series (University Park, Pa.).

HD9000.5.F497 2007
338.1´9—dc22
2007021140

The Pennsylvania State University Press
is a member of the
Association of American University Presses.

It is the policy of The Pennsylvania State University Press to
use acid-free paper. This book is printed on Natures Natural, containing
50% post-consumer waste, and meets the minimum requirements of
American National Standard for Information Sciences—Permanence
of Paper for Printed Library Material, ANSI Z39.48–1992.

CONTENTS

ACKNOWLEDGMENTS

Like the individual, collective, and institutional work required for transforming our agrifood system, this book is the product of a similar collaboration. The momentum for this project originated with the creative energies of members of the International Sociological Association's Research Committee on Agriculture and Food (RC-40). In June 2003 the editors of this volume organized and convened a two-day miniconference in Austin, Texas, entitled "Resistance and Agency in Contemporary Agriculture and Food: Empirical Cases and New Theories." This event brought together scholars and scholar-activists committed to assessing the multifarious changes under way in the agrifood system. We owe much to those at the miniconference for sharing their research and joining us on the path that would ultimately lead to *The Fight Over Food*.

It is almost impossible to recall the many individuals who had a hand in this project from start to finish. From those who encouraged the project to those who offered words of advice along the way, we are indebted. We would like to begin by acknowledging William Friedland and Norman Long, who supported the development of this volume. Special thanks go to Ray Jussaume, Douglas Constance, and Patrick Mooney for their guidance at various stages. We also want to acknowledge Carmen Bain, Alessandro Bonanno, William Friedland, Amy Guptill, Elizabeth Ransom, and Keiko Tanaka for their diligent work in reviewing chapters and providing insightful and constructive commentary.

As editor of the Rural Studies Series, Clare Hinrichs played a key role in encouraging the development of this project and applied her keen eye to numerous drafts. She was especially skilled at helping us fit the conceptual pieces together in the early stages. Peter Potter and Sanford Thatcher at Penn State University Press also proved exceptional in their ability to help us navigate the publishing process, providing much needed feedback and support. We also gratefully acknowledge the comprehensive review of this manuscript by two anonymous reviewers who invested an enormous amount of time and thought in helping us improve the work.

Finally, colleagues and friends are often a sounding board for ideas in the germination stage. We greatly appreciate the collegiality of Kent Sandstrom, Jerry Stockdale, and Robert Schaeffer, who took the time to listen and provided valuable perspective at times when we were mired in detail. This is surely a much improved book because of the efforts of this cadre of supporters, but the editors and contributors alone take responsibility for any omissions, interpretive failings, and conceptual and methodological missteps.

INTRODUCTION

FIGHTING OVER FOOD:
CHANGE IN THE AGRIFOOD SYSTEM

Wynne Wright and Gerad Middendorf

Time magazine recently reported the story of a man in Ann Arbor, Michigan, who had been pulled over in his pickup truck by the state police for hauling illegal cargo. This was the culmination of a sting operation that resulted in seizure of the cargo. But this was no ordinary drug bust; the driver of the mud-splattered pickup truck was a dairy farmer dealing in raw milk (Cole 2007). A growing number of consumers, often from urban locales, are seeking out the warm, white liquid straight from the udder for what they perceive as its superior nutritional value. The Food and Drug Administration (FDA) does not see it that way. The FDA line is that unpasteurized milk contains *E. coli,* salmonella, and listeria—all risks to human health. It is illegal to transport raw milk across state lines, just as it is illegal to sell raw milk in twenty-three states. As a result, farmers and raw milk drinkers have found a creative way to circumvent this obstacle by partnering together to organize a small cooperative venture, known as a cow-share. It is legal to drink straight from your own cow, so by organizing a cow-share co-op, consumers are effectively drinking from the cow they own and partnering with the farmer to do the milking and caretaking. The confiscation of the raw milk has set off something of a maelstrom in Michigan. Raw milk drinkers are a dedicated bunch; members of the farmer's cow-share cooperative could not run fast enough to the aid of their farmer, offering support in numerous ways. It seems that this case is not so unique. People appear to be struggling over the meaning of raw milk—dubbed "real" milk by its supporters—in a number of states. New York, Maryland, Ohio, Kentucky, Indiana, Nebraska, Colorado, Arizona—the raw milk crack down is spreading, while at the same time resistance is mounting. The conflict over raw milk and the growing divide among producers, consumers, and the state is merely the most recent example of a radically changing food system. The case of raw milk suggests that we are increasingly confronting a food and agriculture system[1] that is being restructured in both subtle and highly politicized ways.

In this book we probe the tremendous dynamism of these social forces to explore the lessons they teach us for understanding the complexity of human agency. Some of the agrifood system changes are quiet and almost slip under the radar for their ordinary, everyday properties. Other changes are constantly scrutinized and are highly politicized on the domestic and international stages, bringing a multitude of actors with divergent interests into conflict with one another. We can see the political and sociocultural nature of this process in diverse initiatives, from campaigns to prevent animal cruelty, to the now routine on-site protests against the opening of McDonald's restaurants around the world, to the rise of a new entrepreneurial ethos in producers and a heightened reflexivity among consumers. A casual observer of the media might conclude that something qualitatively new is afoot in humans' relationship to food and agriculture. Consider the following newspaper headlines: "Kinder, Gentler Food: Restaurants, Groceries, and Activists Force Farmers to Change Treatment of Their Animals" (Perkins 2004); "Monsanto Lab in Crystal Closes Amid Food Protests" (*Bangor Daily News* 2000); "The Slow Food Movement: Slow Food Protests over Fast Foods and Food Safety" (*New Internationalist March* 2002); "Tests Confirm 2nd Case of Mad Cow Disease in U.S." (Newman 2005). If headlines are barometers of social trends, then it appears that we are indeed witnessing an explosion of efforts aimed at reconfiguring our relationship to agriculture and food.

We are experiencing the rise of an "alternative" food system that attempts to exist outside of the mainstream commodity-driven network. This alternative network comprises a repertoire vast in economic scale, political intentions, and cultural overhaul. Growth in interest and activity around organic foods, eco-labeled foods, direct marketing, fair trade, local foods, community kitchens and gardens, community-supported agriculture, food box schemes, farmers' markets, and assorted community buying clubs collectively demonstrate the emergence of a new production/consumption paradigm and are some of the topics taken up in this book. The resurgence of such locally embedded food networks is growing in the United States, but Americans are not alone in charting a new food course. Whatmore, Stassart, and Renting (2003, 389) conclude that such projects "represent some of the most rapidly expanding food markets in Europe over the last decade." But efforts to

---

1. Hereafter, we abbreviate the food and agriculture system "agrifood system."

construct new alternatives to the existing food system are not the only subject we explore in this book. At the same time, challenges to the current commodity-driven food system by producers and consumers are growing. Struggles over the organization of the current food system are evident in food antidisparagement laws, as well as in the backlash against confined animal feeding operations (CAFOS), aquatic "feedlots," genetically engineered foods, mad cow disease, *E. coli* contamination, biopiracy, and dissent targeted at transnational trade regulatory bodies such as the World Trade Organization (WTO).

These efforts are only some of the phenomenal changes taking place in the agrifood system. A complete list might be nearly impossible to compile. Therefore, we have been somewhat selective in our account. The essays in this volume highlight some of the major transformations with which many readers will be familiar and, at the same time, demonstrate the arenas of contest that are increasingly part and parcel of our food system.

It is this broad arena of contest—the aggregation of efforts to reshape the conventional agrifood system—that frames our conceptualization of the fight over food. What do these streams of social change share in common? They signify a mounting reflexivity and new modes of action among producers, consumers, and activists in the production and consumption of food (Fine 1995; Fine and Leopold 1993; Goodman 1999, 2004; Lockie et al. 2002; Murdoch, Marsden, and Banks 2000). Food, along with its attendant production processes, is moving to the forefront of our consciousness. It is being reconsidered in light of changing values, norms, customs, science and technology, and institutional resources, and is no longer invisible in our culture. Many of our long-held assumptions about food—from the way it is produced to the way we eat—are now in flux.

These new configurations provide us with an opportunity for a sociological study of the various ways that agency is expressed by humans. All of these contests in the agrifood system—from the rising demand for raw milk to the segmentation of our local supermarkets—are products of human agency. We recognize that social structure can direct, even constrain agency, yet we have avoided overly structuralist accounts that suggest that agency can be squelched under the unbending rigor of structure. We also steer away from interpretations that seem to imply that agency is free to be realized without any constraint whatsoever. In the transformations taking place in the agrifood system we are reminded that actors exercise agency in numerous ways. They may reflexively and creatively generate new forms of action (e.g., new markets), or they may resist the establishment of new forms of action or the continuity of activity. Finally, this volume examines the ways in which individual action and collective activity are articulated. It links the reflexivity of consumers with the collective acts of citizens and attempts to prompt a dialogue that focuses on the synthesis of these roles along with other social roles humans play in society.

How did we get here? In the postwar era of abundance, food moved to the back burner of the consciousness of many in the industrialized world. For most, it became plentiful, inexpensive, more convenient, and perceived as relatively nutritious. Given a willingness to trust those embedded in our food-provisioning system, we relinquished our civic responsibility for food system oversight to farmers, nutritionists, food corporations, agribusinesses, and the state; in other words, we let the experts take charge. Issues of how, where, and by whom food was grown were not generally topics of conversation around the dinner table. In some circles, it might even be considered unacceptable or impolite to inquire about the social life of our dinner.

As we migrated away from the farm and the dinner table through the twentieth century, our consciousness of food likewise migrated away from the biological and social basis of production. As women moved into the workplace and those in the West generally increased the hours spent at work outside the household each week, interest increased in processed foods that provided convenience (e.g., canned, frozen, and prepared foods) (Goodman and Redclift 1991). All the while, confidence in the food system soared— we were assured that food processors were providing us with nutritious choices and that food scientists were using state-of-the art technology to resolve food safety issues. The choices available to the typical shopper appeared to multiply exponentially on supermarket shelves. Most of us believed that our food was being produced by farm families who received a living wage for their labor. Their dependence upon healthy soil and clean water assured us that they would not compromise environmental integrity through poor production practices. Yet we now know that the story of postwar development of the agrifood system is one of both successes and failures.

Although initially it was ridiculed, Rachel Carson's *Silent Spring* (1962) shocked many out of their quiescence over the environmental impacts of indiscriminate pesticide use in agriculture. Since that time, numerous scholars have demonstrated the role agricultural policy has played in encouraging a production treadmill that results in relatively inexpensive and abundant foods, while at the same time expelling farmers from the land, with crippling effects for rural communities (Brown 1988; Cochrane 1979; Dudley 2000; Strange 1988). And now, in the early twenty-first century, we have turned our attention to issues of health and safety (Critser 2003, Nestle 2002). As it turns out, the glorious choices that literally bulge from the shelves of grocery retailers also symbolize the price many in the advanced industrial world pay for a diet of affluence.

These growing realizations stem, in part, from our postmaterialist age, in which some of us have the luxury of putting quality and identity issues surrounding food front and center in our consciousness. But many in less developed countries, as well as an embarrassing number in industrialized nations, continue to combat issues of food insecurity with unfortunate regularity. Ten years after the World Food Summit, where we pledged to tackle the problem of hunger and food insecurity, "virtually no progress has been made." The Food and Agriculture Organization of the United Nations estimates that 854 million people suffer from undernourishment worldwide, 845 million of them in developing and transitional economies and 9 million in the industrialized world (FAO 2006). In the Unites States alone, 11.9 percent of the population—or about 35.7 million—is food insecure (Nord, Andrews, and Carlson 2004). The simultaneous proliferation of food boutiques for the wealthy and food banks for the poor is a distressing paradox that defines an era of overindulgence alongside deprivation (Van Esterik 2005).

What interests us in this volume are the numerous and multifaceted initiatives in the agrifood system that attempt to bring about change in some way, whether through an effort to address social disparities such as hunger and food insecurity or to raise health or environmental concerns, change our eating patterns, or diminish corporate influence. Food and agriculture have become public debates in transnational, national, and regional policy, in part because they offer what social movement theorists refer to as a "consensus frame." The desire for accessible "quality" food, a healthy environment, and regional economic development transcends the social markers of race, class, gender, and geography that divide us. We all want and need sustainable livelihoods.

Social scientists have variously referred to this distinctive impulse in food provisioning as an "alternative geography of food" (Whatmore and Thorne 1997) or an "alternative agro-food network" (Murdoch, Marsden, and Banks 2000). We find particularly useful Lyson's concept of "civic agriculture" as a way of describing this remaking of our agrifood system. Civic agriculture is an organizational production strategy "tightly linked to a community's social and economic development" (Lyson 2004, 1). This approach offers an integrative vision that promotes local environmental stewardship and rural community culture, including reciprocity and norms of neighboring, while simultaneously incubating engines for economic development. In this way, it liberates communities and food systems from global dependence by promoting paths of local interdependence. We feel that Lyson's concept of civic agriculture comes closest to reuniting us with our food and our social obligations to each other. We will return to this subject in the concluding chapter. In this volume, we use a case study approach to analyze the developmental processes of food system change, thereby filling a gap in the agrifood literature.

We examine many of these trends and ask whether, taken together, they portend an "accumulation of resistance" that may fundamentally transform the predominant agrifood system, or whether they represent rather fragmented, atomized expressions of symbolic consumption. In other words, do these trends signify a transformative social movement or merely the emergence of "bourgeois bohemian" or "bobo" (Brooks 2001) food consumption patterns? Have the affluent classes merely reembraced food as a form of social capital (Bourdieu 1984), or do the efforts to redesign our food system suggest the birth of a new institutional arrangement, one that could replace the conventional productionist-centered agrifood system?

While anthropologists, economists, geographers, and agricultural sociologists have tried to present the range of new agricultural and food initiatives and explain their resonance, few have attempted to draw them together and examine their transformative potential. Seldom have we seen in the social science agrifood literature cogent attempts to grapple with these initiatives as a comprehensive form of social change. Can they be explained as a function of new forms of human agency, or as resistance to the dominant mode of agriculture? Or both? In this volume, we explore these questions.

An exhaustive review of the literature is beyond the scope of this introduction, but we hope to provide some tools to help the reader grasp the ways in which social change is possible. We turn first to an overview of some of the changes taking place in our agrifood system. In other words, we look at what's "new" in agrifood systems. This involves the discussion of a concept of paramount interest in this volume—human agency. We provide an overview of this concept and touch briefly on some of the conceptual debates in the literature where they relate to agrifood system change. We conclude the introduction with a brief overview of the case studies in the book.

The Politics of the Plate

Starting with western Europe, van der Ploeg and Renting highlight the invigoration of an entrepreneurial ethos and demonstrate the salience of new critical market streams for the production sector. They write that more than half of professional farmers in the European Union (EU) have adopted on-farm production strategies that add value "through such mechanisms as quality production, on-farm processing, marketing through new short circuits," or are attempting to broaden their market reach by expanding into "new non-agricultural activities that are located on the interface between society, community, landscape and biodiversity" (van der Ploeg and Renting 2004, 235). Organic agriculture is one example of such "quality" production.

ORGANICS

Fast becoming one of the most dramatic expressions of the politicization of food, the popularity of organic food and beverages has accelerated rapidly. The Food and Agriculture Organization (FAO) and the World Health Organization (WHO) define organic agriculture as a "holistic production management system which promotes and enhances agro-ecosystem health, including biodiversity, biological cycles, and soil biological activity. It emphasizes the use of management practices in preference to the use of off-farm inputs, taking into account that regional conditions require locally-adapted systems. This is accomplished by using, where possible, agronomic, biological and mechanical methods, as opposed to using synthetic materials, to fulfill any specific function within the system" (Slign and Christman 2003, 1). The International Federation of Organic Agriculture Movements (IFOAM) goes further by integrating components of social and economic sustainability, along with biological elements, into its definition of organic agriculture (IFOAM 2007). Regardless of differing definitions, it seems that organic agriculture has exploded on the world stage, affecting production and consumption trends and creating new industries as well as reconfiguring older agribusiness firms (DuPuis 2000).

Global sales of organic products are increasing annually, making organics a growing segment of the food market. In the United States, organic production has outpaced all other sectors of the farm economy (OTA 1999), with sales growing by 20 to 25 percent annually over the past decade (Greene and Dimitri 2003). This surprising growth in the U.S. demand for organics has bypassed that of the European Union, historically the leader in organic consumption (*Organic Monitor* 2006). More than 77 million acres worldwide are now farmed using organic methods (IFOAM 2007), 4 million of them in the United States. The economic future of organic production appears bright in many areas of the world. Organic food and drink sales in the United States hit $40 billion in 2006, and many areas of the world began to experience shortages (*Organic Monitor* 2006). But an expanding market share does not readily translate into the potential to fundamentally transform our agrifood system.

It is precisely this market momentum that has caught the attention of the purveyors of industrial agriculture who want a piece of this fast-growing pie for themselves. In a study of the California organic sector, Guthman (2004) found that agribusiness firms were rapidly capturing a vast share of the organic market and applying their familiar production-intensive and industrial template to the sector. This control, Guthman argues, translated into a separation of organic production from its philosophical foundation—historically in opposition to the capitalist mode of production. Instead, the organic sector is coming increasingly to resemble other sectors of commodity-driven agriculture, and in the process fraying the oppositional threads necessary for transformative change.

In addition, there remain questions about the diffusion of organic products to the general public. While growth of this new sector has been robust and while it is becoming increasingly standardized, consumption continues to be influenced by social class and may be out of the economic or cultural reach of the working and lower classes (Ehrenrich 1985; Friedland 1994; Kauffman 1991; Marsden and Arce 1995). Even among the middle class, there appears to persist a habit or custom of cost saving, making the purchase of organic and other high-end products seem frivolous or unnecessary when foods perceived to be sufficient and comparable can be purchased less expensively at chain stores like Sainsbury's, Tesco, Loblaws, and Wal-Mart. At the moment, it appears that organic food and beverage production has succeeded largely by crafting an upscale market clientele willing to pay premium prices for either quality or status, to the delight of supermarket retailers who are scurrying to capitalize on this market segmentation. Miele (2001) found that European consumers are willing to pay high prices for organic products, ranging from 20 to 200 percent above the cost of nonorganic fare. In other regions, producers have been slow to realize the market potential of organics. In Chapter 8, Amy Guptill looks at the paradoxical case of Puerto Rico, which has deftly organized an alternative agrifood infrastructure (markets and grassroots activism) but faces a legacy of social and economic arrangements that present a formidable challenge to organizing an effective production sector.

LOCAL AND REGIONAL FOODS: FARMERS' MARKETS AND CSAS

One of the early market outlets for organic products was local venues that synthesized community, place, and market exchanges, such as community-supported agriculture and farmers' markets and other direct-market schemes, and, in the process, bypassed the global industrialized food system (Henderson 2000). Local food systems are "rooted in particular places, aim to be economically viable for farmers and consumers, use ecologically sound production and distribution practices and enhance social equity and democracy for all members of the community" (Feenstra 1997, 28).

Farmers' markets have been viewed as the centerpiece of food system localization efforts, acting as dynamic venues that serve multiple functions: diversifying homogenous food retail outlets, channeling revenue to cash-strapped producers, incubating small business development, and reembedding local producers and consumers in market exchanges and community-building networks (Hinrichs, Gillespie, and Feenstra 2004, 32). It is not surprising that they would figure so prominently in food system renewal, given their roots in social resistance. Brown recounts the revival of California farmers' markets in the early 1940s as an illustration of organizers' pragmatic response to postwar produce-distribution problems. The 1940s saw the first American renaissance of farmers' markets; the second took place in the late 1970s. Proponents of local markets faced a groundswell of political opposition during the authorization of the Farmer-to-Consumer Direct Marketing Act of 1976 (Public Law 94-463), which charged that "direct marketing threatened the national food supply" (Brown 2001, 669). Regardless of the social and political obstacles, their resurgence in the United States has been nothing short of phenomenal, increasing almost ten times over the past three decades. Even more impressive is the growth that has occurred in the past decade. Farmers' markets in the United States grew from 1,755 in 1994 to 4,385 in 2006, a growth rate of 150 percent in twelve years (USDA 2006).

Farmers and consumers alike appear to be drawn to open-air retail markets for a number of reasons (Lyson, Gillespie, and Hilchey 1995). According to Hinrichs, Gillespie, and Feenstra (2004, 34), the resurgence of farmers' markets in the 1990s is due to "producers' renewed search for more profitable alternatives to wholesale commodity markets, consumers' rising interest in farm fresh and regional specialty foods, and also the cachet of colorful open-air markets as trendy arenas for consumption." The popularity of some markets, such as the Dane County Farmers' Market in Madison, Wisconsin, demonstrates that, for some, the market itself has become a destination point, drawing community members and tourists to a wide panoply of festive, family-centered events. Many markets offer a convivial atmosphere and in this way provide the foundational elements for community cohesion as they generate an atmosphere of cultural celebration, whether through neighborly interaction, petting zoos and face painting for children, or community string quartets that provide a highbrow backdrop to the consumption experience. Andreatta and Wickliffe (2002, 6), in their survey of market patrons in North Carolina, found that 15 percent reported that a trip to the local market "was something fun to do on a Saturday morning." While local, they can also nurture a global culture (Waterman 1998) through the presence of new and ethnic foods. A stroll down the aisle at some large metropolitan markets reveals the infusion of new Central American, eastern European, African, and Asian foods. In the same way that organic products tend to be largely the domain of the more affluent, farmers' markets in many areas have also become the shopping destination for those who have high levels of disposal income, demonstrate variable commitment to the politics of alternative food systems (Hayes and Milánkovics 2001), and live in urban rather than rural areas (Hinrichs, Gillespie, and Feenstra 2004).

While the popularity of farmers' markets soars, community-supported agriculture (CSA) schemes have also demonstrated impressive potential to re-embed markets in localities by directly reconnecting producers and consumers. CSA is a unique consumer-producer partnership that allows both parties to share in the risks and rewards of food production. Subscribers (consumers) pay farmers in advance for a share of fresh produce that is delivered to them regularly during the growing season. Farmers benefit from the locally based direct markets for their produce and access to investment capital from subscribers to offset production expenses. Consumers have the chance to reconnect with the producers of their food and gain access to high-quality fresh produce. They are also able to enhance their knowledge of the production process and of farming and environmental issues more generally. CSA shareholders often speak of the new varieties of vegetables to which they are exposed and their education in rural and agri-environmental politics. Both groups benefit, as they are able to nurture bonds of affinity. As one Kentucky CSA operator put it, "I don't have seventy-five shareholders, I have seventy-five friends."

One of the key differences between CSA and farmers' markets is the level of commitment. CSAs generally involve a higher level of commitment on the part of the participants than do farmers' markets, as they require subscribers to prepay for seasonal produce rather than frequent a market episodically (Hayes and Milánkovics 2001). Moreover, CSA arrangements vary widely, depending on the objectives, needs, and preferences of both the farmers and the subscribers. Depending on labor needs and philosophy, the arrangement may require a minimum of a prepaid subscription that farmers use to cover the costs of production for a season. Membership may also involve volunteer time on the farm, however, or a more formalized "working share" in which subscription costs are partially offset by members' labor. This may explain in part why the CSA model has not only been successful in producing fruits and vegetables but has also been widely touted for its ability to foster community (Andreatta and Wickliffe 2002; Groh and McFadden 1997; Lyson, Gillespie, and Hilchey 1995; Hinrichs 2000). Since the introduction of the CSA model in the United States in the mid-1980s, it has expanded to more than a thousand operations in North America (Hendrickson 1999).

FAIR TRADE: EQUITY IN FOOD?

Others are looking to Fair Trade networks to make more overt political commentary about the equity of food distribution. Raynolds (2002, 410), for example, writes that Fair Trade "involves the constitution of alternative knowledge systems as well as commodity networks," by challenging the inequitable organization of conventional world markets and attempting to repair the exploitive relations embedded therein.

Thus far, Fair Trade networks tend to include tropical commodities like coffee, tea, cocoa, and bananas. Whatever the product, Fair Trade networks attempt to "shorten the social distance between consumers and producers" and, based on trust and a sense of justice, offer farmers a more equitable piece of the food dollar, sharing civic and ecological responsibility with consumers (Raynolds 2002, 420). According to Raynolds, the Fair Trade market has grown to a value of $400 million. Not unlike organic food production, farmers' markets, and CSAs, Fair Trade networks continue to enroll impressive numbers of actors, with North American and Pacific Rim sales growing by 37 percent in 2002 alone. Clearly this market has grown in scale, but does this growth translate into improved living conditions for those who rely on it? A related question is whether Fair Trade poses a challenge to conventional market streams. Aimee Shreck (Chapter 5) examines the transformative potential of Fair Trade by exploring conditions in the Dominican Republic.

Challenges to Agriculture as Usual: CAFOS and Biotechnology

CAFOS

These examples of agrifood system change are coupled with other struggles that attempt to challenge the detrimental effects associated with conventional models. Among the most insidious reminders of the persistent productivist bias associated with highly rationalized systems of food production are the environmental and social impacts of confined animal feeding operations (CAFOs) that have inspired individual and collective mobilization (Bonanno and Constance 2000; DeLind 1998; Thu and Durrenberger 1998; Wright 2004). Community after community in the United States has seen protests over the siting of CAFOs in its vicinity, including floating feedlots or aquatic CAFOs in some fishing communities. Canadian activists declared a "day of action" against intensive animal production facilities in April 2004, charging that the primacy of economic interests in rural communities has led to threats to human health, environmental well-being, and community cohesion (Brubaker 2004). Americans and Canadians are not the only ones to cry foul. Polish activists successfully defeated an effort by Smithfield Foods, the world's leading pork producer, to transplant its industrial-style hog production into Poland's fragile transitional economy. Success was achieved by a partnership between the Animal Welfare Institute of Washington, D.C., Polish farm groups, government officials, and representatives of the Polish press (Juska and Edwards 2004). This international partnership is a good example of the growing transnationalization of agrifood movement actors who are learning that their futures are intricately bound together. Grassroots organizations and community groups have called for moratoriums on the construction of new CAFOs to thwart industry expansion and halt the reshaping of the landscape into a "rural ghetto" (Davidson 1996).

The issues associated with CAFOs are familiar to most students of agriculture. Opponents contend that intensive livestock operations constitute a rural "race to the bottom," both culturally and economically, as they cause respiratory problems among employees and health ailments for those living in the vicinity of the operation, pollute water sources from runoff, lagoon spills, and leakage, and degrade neighbors' quality of life owing to the odors, water pollution, and community contention they create (Jackson 1998; Kleiner and Constance 1998). Most of the complaints are directed at the operations on an individual level, but some challenge decisions that place disproportionate authority in the hands of nonlocal residents. Others weave issues of animal welfare into their concerns about this production strategy. The growing recognition of the social problems associated with CAFOs has encouraged some producers to pursue alternative production strategies, such as methods that allow meat to be produced in a more socially and economically sustainable manner (e.g., Swedish deep-bedding systems, hoop structures, open grazing, and community cooperatives).

Agricultural Biotechnology

No discussion of agricultural change would be complete without addressing the contentious political struggle of the past two decades over agricultural biotechnology. Since the approval of the first products of biotechnology in the food system in the early 1990s, debates have raged over the economic, social, ethical, and environmental consequences of the "biorevolution" in agriculture. Early debates about the actual impacts of the large-scale use of biotechnology tended to be speculative, because its products had not yet been planted in large scale or commercialized for public consumption. Yet much was already under way, and scholarship at the time focused on industry-university relationships (e.g., Kenney 1986), the role of biotechnology in the further industrialization of agriculture (Goodman, Sorj, and Wilkinson 1987; Kloppenburg 1988), patenting and intellectual property rights (e.g, Busch et al. 1991), and the role of developing countries in biotech development (Souza Silva 1994), among other issues. The debate heated up substantially with the highly publicized approval, in 1993, of recombinant bovine somatotropin (also known as rBGH, a recombinant bovine growth hormone injected into milk cows to increase production) and Calgene's "Flavr Savr" tomato in 1994. The approval of genetically engineered foods for market brought increased attention to a new host of issues, such as the ethical implications of food biotechnology (Thompson 1996, 1997), including labeling and consumers' right to know (e.g., Guthman 2003). Opposition groups became more active, diverse, organized, and transnational. They have organized conventional protest events around the world but have also developed and disseminated information, targeted retailers, affected trade meetings, and probably influenced consumer patterns (Munro and Schurman, Chapter 6 in this volume).

During the mid- to late 1990s, a host of other genetically engineered products were approved for release. Perhaps most prominent among these were seeds that had been engineered with one of two key traits: herbicide resistance (e.g., Roundup Ready soybeans, canola, corn), or insect resistance, achieved through engineering seeds to contain a protein from the soil microbe *Bacillus thuringiensis* (Bt) (e.g, Bt corn, cotton). Since the release of these seeds, the global area planted in transgenic crops has increased dramatically, from about 11 million hectares in 1997 to 40 million in 1999, then doubling to about 81 million hectares globally in 2004 (James 1998). While the global planting of transgenic crops has now diversified somewhat by country, crops, and traits, the vast majority of this acreage remains in three countries (the United States, Argentina, and Canada account for 85 percent of global acreage), two crops (corn and soybeans), and the two traits mentioned above. The expansion of transgenic acreage planted has been accompanied by increased research on the environmental implications of agricultural biotechnology (e.g., Krimsky and Wrubel 1996; Rissler and Mellon 1996; Mikkelsen, Andersen, and Jørgensen 1996; Altieri 2004).

In recent years, one of the key themes in the debate has been global trade issues. The biotech industry has argued that genetically modified crops and foods are essentially equivalent to nongenetically modified foods, and thus should flow freely in global trade. Yet there have been numerous moratoriums or bans on United States grains by importing countries that have chosen a more cautious approach to these grains and food products. Moreover, at global trade meetings, dissent against genetically modified crops and foods is one of the themes taken up by antiglobalization groups as well as spokespersons for developing countries. The public protests at the WTO meeting in Seattle in 1999 are emblematic of this new form of resistance to agricultural biotechnology. In Chapter 6, Munro and Schurman examine the origins and historical development of the anti–genetic engineering movement in the United States and attempt to assess its successes and limitations as a collective agent for social change.

These trends are only a few examples of the new socioeconomic and political repertoire of those engaged in the remaking of the agrifood system. This inventory is by no means exhaustive, but we hope it will suffice to give the reader a sense of the breadth of these initiatives. These actions, and others like them, raise the question of how agency is being manifested in agrifood systems and the ways in which the exercise of agency is constrained.

"The Food at the Center of the Plate": Agency

Agency and structure are pivotal concepts in contemporary social theory, and many theorists have attempted to define, distinguish, and integrate the terms (see Archer 1982; Bourdieu 1984; Giddens 1984; Habermas 1987).[2] As well they should. The examination of agency and structure in contemporary agrifood systems is an important sociological inquiry, because it sheds light on how humans shape something as essential for life as food, and on how the existing food system shapes human action. We should clarify from

---

2. The heading of this section is an adaptation of an earlier critique of Giddens's work on structuration theory (Craib 1992, 196).

the outset that we see agency as active, reflexive choice that is embodied in either individuals or collectivities (Burns and Flam 1986). Agency is the ability of humans to act purposively, of their own volition, and to some extent independently of the constraining aspects of structure, including the predominant customs and norms of culture. For example, individuals may effect change by switching from conventional to organic milk (DuPuis 2000). This is a form of agency. Others might see it as a form of everyday resistance (Scott 1985), yet resistance, whether undertaken by an individual or a large group, is an exercise in action in the face of perceived structural conditions that may limit or detour agency.

Altering your habits of milk consumption, or of any consumable good for that matter, while personally significant, is localized change that affects only the individual or, at best, the household. It is unlikely to bring about wider transformative change unless diffused to a broader audience that has the power to effect change through the power of numbers. In the 1960s, César Chávez, California activist for farm workers rights, succeeding in mobilizing thousands to boycott grapes until grape growers responded by improving worker wages (Mooney and Majka 1995). These examples signify two kinds of agency. When we reflexively choose alternative foods, we demonstrate agency, but agency comes in other shapes as well. Large numbers of individuals working collectively, like Chávez and the dedicated farm workers, may communicate their grievances and effect change by organizing consumer boycotts. Mounting crop-withholding strikes, barricading roads, or dumping milk and butter, as Iowa farmers did during the Great Depression in response to the harsh effects of monopoly capitalism (Mooney and Majka 1995), are other instances of agrifood agency. Collective action stands a better chance of realizing systemic change, but it is by no means a silver bullet. Agency invariably runs up against obstacles of structure, yet it is important to recognize that humans, in the exercise of agency, are in a continual process of reshaping those structures to varying degrees.

There are a number of ways to conceptualize structure. Some scholars have emphasized that social life is largely determined by structure and that individual agency is explained by the prevailing structures. In this view, structure is often seen as residing outside the individual and constraining human action, or as the backdrop that determines the social activity taking place in the foreground. Others have described structure as consisting of patterned forms of organization and culture that are embedded in society (Ritzer 1996). Still others have emphasized structure as an outcome of social relations and interactions—for example, structure as a nexus of congealed action (Booth 1994), as negotiated order (Busch 1980), as a layering of social relations over time (Giddens 1984), or as stabilized networks (Latour 1987). These latter versions tend to open up more space for agency, for if structures are an outcome of social relations, then they can be changed through human agents acting reflexively and purposively. The point is that this is always a dialogical process in which structure is both medium and outcome. As Giddens (1984, 25) argues, structure is always both constraining and enabling. Culture and social organization may limit or close opportunities for food system transformation; in other cases, culture and embedded organizational patterns may facilitate change. Perhaps one of the clearest examples we can provide to illuminate what might seem like a paradox comes from a lesson taken from American tobacco farmers.

In a class action in 1998, the federal government required tobacco manufacturers to pay compensation of $206 billion to forty-six states and six territories to cover medical treatment for those with tobacco-related illnesses. This lawsuit is known as the National Tobacco Settlement. In an effort to offset their economic loss, transnational tobacco firms responded by purchasing their tobacco leaf from lower-cost production regions (e.g., South America and Africa), which allowed them to reduce their dependence on the higher-priced leaf from American producers. The economic consequence for farm families and tobacco-dependent communities has been a loss of reliable streams of rural revenue.

As the structure of tobacco production changed following the lawsuit verdict and the decline in national markets, opportunities that farmers had previously enjoyed were closed off. But embedded in this new structural obstacle was a new opportunity. In light of the hardship tobacco farmers were expected to face, legislators in states with a large tobacco-producing population allocated some of the financial windfall to agricultural diversification and rural revitalization. The Kentucky General Assembly appropriated more money than any other state for such efforts. Set to receive $3.45 billion over the next twenty-five years (Hall, Snell, and Infanger 2000), the state created an unprecedented opportunity to prioritize rural economic and social development. Assistance in the form of grants and low-interest loans has allowed farmers who had feared for their economic future to exercise new forms of agency as they go about the work of transitioning away from tobacco and building a more sustainable agrifood system. In this case, the sociopolitical reorganization of tobacco production created structural obstacles but at the same time provided a window of opportunity for farmers to express new forms of agency.

This story reminds us how structure can constrain and enable at the same time, yet it is not complete. Kentucky farmers and other grant seekers are not free to chart whatever agricultural future they might desire. They are obliged to adhere to the rules and procedures for grant seeking and other support based on predetermined regulations developed by a specially appointed Agricultural Development Board (Eaton 2004). Candidates seeking funds for tobacco transition are required to meet a number of criteria that fulfill the board's definition of model agriculture.

As the case of tobacco and the other case studies in this volume will make clear, the line where agency begins and structure ends is quite blurry—the two things are part of a duality and cannot be neatly separated (Giddens 1984). For example, while we see the development of new market streams, such as organic production, as an exercise in agency, we also view this in some cases as a backlash, or a form of resistance to commercial market streams that squeeze small producers to the point of pushing them out of the marketplace altogether. Producers resist expulsion from the market at the same time that they creatively act to create new niche markets. Rather than seeing agency and structure as part of a dualism (see Archer 1982), we suggest that these cases be read with an eye toward understanding these complex concepts as dueling tensions that alter form and substance given changing contexts. We are reminded of Marx's famous dictum, "men make history, but they do not make it just as they please; they do not make it under circumstances chosen by themselves, but under circumstances directly encountered, given, and transmitted from the past" (Marx 1869/1963, 15). It is this dialectical tension that we hope readers will appreciate in the cases presented in this volume, as we search for the instances where human agency is transforming the agrifood system. We return to this friction in the conclusion and ask if such contradictory tensions can be useful.

Overview of the Book

We have organized this volume into three sections that are intended to give the reader a glimpse into major changes taking place in the agrifood system. Agency is the linchpin, or the primary sociological concept, around which this book is constructed because of its essential quality in realizing the transformative potential of agrifood system renewal. Many of the chapters commingle the concept of agency with equally illustrative concepts, such as resistance and structure. The vastness and complexity that characterize the remaking of the agrifood system prevents us from providing an exhaustive treatment of the restructuring efforts. It is our hope that readers will take from the following chapters an appreciation of this phenomenon and the hunger to discern the changes taking place in their local supermarket, farming community, restaurant, farmers' market, and food pantry.

The first section is devoted to three essays that provide either the historical context or a theoretical framework for understanding agrifood system change. The case studies, which reveal that human agency is alive and well, are concentrated in Parts II and III. We can discern two significant empirical strands in these case studies. First, agrifood system change is a highly variable and uneven process. In some instances we find actors reflexively acting to transform the agrifood system by altering their eating patterns; in other cases, change is occurring on a larger scale, such as sweeping market reorganization. Second, most if not all of the essays demonstrate contradictions and tensions within the agrifood system, suggesting that formidable structural hurdles must be overcome if lasting change is to be realized. Given these two conclusions, we have organized the case studies around the relative consequences of agency and structure in assessing the potential for agrifood system transformation. In distinguishing between agency and structure, Ritzer (1996, 560) argues that "the real issue is not agency and structure per se but the relative weight of agency and structure." He charges social analysts to examine agency and structure on a case-specific basis and avoid treating them as equivalent across time and space.

Because of the observable differences in the exercise of agency among the actors represented in these case studies, the section of the book devoted to agency is organized in two parts. Part II, "Making Room for Agency," examines cases that show us that the exercise of agency appears to be bringing about change. Readers will discover that the form of agency may vary somewhat. For instance, agency may be observed on an individual, collective, or organizational level. It may be responsible for subtle or incremental change, or it may present an overt political challenge to the status quo. Nonetheless, the case studies in this section show the greatest potential for the exercise of agency and change.

The case studies in Part III are characterized by more obstacles to the effective exercise of agency than those in the previous section. These obstacles make change slow and labored. This part, "Constraints to Agency," highlights case studies that reveal difficulty or in which actors have been unable to identify or construct favorable socioeconomic and political outcomes. Yet these cases are nonetheless cases of agency. Of paramount importance here is the role played by structure and culture, which can restrict actors in their pursuit of changes in the food system. The purpose of making these distinctions is to signify that agency and structure are always part of an ongoing dialogue. All of the case studies probe the dimensions of agency in an effort to reveal the presence of (or constraints to) agency for bringing about transformative change.

While we feel that activists can glean many useful concepts for social action from this book, it is not intended as a "how-to" manual. The essays included here illuminate what has succeeded in various struggles for social change and what forces have proved to be obstacles to the development of a model agrifood system.

Part I of the book, "Conceptual Framework," attempts to unravel some of the historical, conceptual, and interpretative problems implicated in studying issues of agency as they apply to agriculture and food. Alessandro Bonanno and Douglas Constance open the volume with a theoretical and historical roadmap of sorts. They employ an organizational profile of the International Sociological Association's Research Committee on Agriculture and Food (RC-40) to give the reader a background in this scholarship. Readers new to this material may find this background a helpful guide to locating theoretical works. Others may find the organizational detail appropriate for readers more familiar with the subject. Bonanno and Constance round out their chapter by summarizing the key themes addressed in the scholarship of those affiliated with RC-40.

The strength of this opening chapter is the authors' ability to deftly compare and contrast the two chapters that follow, written by William H. Friedland and Norman Long, respectively. These chapters serve as models of the theoretical frameworks that have predominated in the sociology of agriculture. They offer insightful context and prepare the reader to explore the neo-Marxian and actor-oriented approaches that have usefully invigorated debate and enhanced our understanding of agrifood system restructuring.

Relying on a neo-Marxist framework, Friedland characterizes agency as a manifestation of resistance to aspects of capitalist or authoritarian life, as people seek to assert some control over their lives. Manifestations of agency, however, take many forms, some more coherent than others, some challenging authority directly, others indirectly. Several chapters in this book share a similar conceptualization of resistance as a form of agency. Johnston, Shreck, and Munro and Schurman all present provocative case studies that demonstrate resistance, in varying forms, to hegemonic capitalism and its alienating consequences.

Long takes a contrasting approach when he proposes a return to an understanding of the fundamental sources and trajectories of change and heterogeneity, but without class categories or other prewritten scripts, such as globalization. Here, too, the reader will find complementary chapters in this book that operate under the same theoretical predisposition. The chapters by Jussaume and Kondoh and Skladany present empirical cases that forego macro templates and adopt an actor-centered orientation.

Part II begins with case studies that give the reader a glimpse into the wealth of cases that demonstrate agency in the agrifood system. We begin with Josée Johnston's analysis of a Canadian-based food box scheme. She distinguishes between meaningful structural change and "tofu politics" and asks whether the consumption of organic food, support for farmers' markets, or dining at upscale organic restaurants such as Chez Panisse are merely forms of "bourgeois piggery." By examining a nonprofit community food security organization in Toronto, she shows us how certain food projects can be self-indulgent and exploitive, while others work to subvert social and ecological exploitation. We then move to Aimee Shreck's intriguing analysis of the Fair Trade banana initiative. The Fair Trade movement, Shreck argues, has little transformative potential at present. While Fair Trade has been successful in enabling consumers and producers to commit acts of resistance and in facilitating the redistribution of resources from the global North to the global South, she asks whether the movement has yet to attain its full oppositional promise.

Turning more directly to collective action networks, William A. Munro and Rachel A. Schurman's analysis of the anti–genetic engineering movement gives us a fascinating glimpse into the postmaterialist values driving protests against the agrifood system. For these authors, agency is manifested as activism, or resistance to the purveyors of industrial agrifood. Through a social movement lens, the authors explain the origins of these protests and the inroads that are being made by anti-GMO adherents. In the final chapter of Part II, Mike Skladany attributes agency to salmon technology and science when he finds that struggle over salmon is a highly contested ordering of salmon, human, and nonhuman actors. Like Munro and Schurman, Skladany finds signs of hope in the collective action networks of salmon advocates.

The case studies in Part III address less successful inroads into reforming the agrifood system. The chapters in this section deliver informative lessons about the difficulty of exercising agency in transforming the agrifood system. Amy Guptill launches this section by asking why Puerto Rico has watched other areas of the Caribbean invest in organic production yet has not done so itself. She suggests that the answer is that agency manifests itself in the form of resistance exercised by a powerful business class that promotes the consumption of imported goods over local organic production, often employing political manipulation and outright violence to achieve these ends. Such dependence on imports functions as a formidable obstacle to moving toward a sustainable food system.

Raymond A. Jussaume and Kazumi Kondoh take us to the Pacific Northwest to examine the extent to which a consensus may be emerging between consumers and farmers in response to the organization of new agrifood systems. For these authors, agency is manifest on the part of producers as they move into direct-marketing arrangements, but this is often a response to structural obstacles found in conventional channels. Analyzing the differences in *how* producers and consumers are participating in emerging direct-marketing activities, they conclude that structural forces are mediated by social, demographic, historical, and ecological factors that are unique to each locality.

The final chapter, by Keiko Tanaka and Elizabeth Ransom, conceptualizes agency as a network effect. By focusing on changes in the food safety regulatory framework in the New Zealand and South African red meat commodity chains, they show how multiple agencies are constructed and implicated in the red meat chain to justify economic and political interests.

We have included a number of resources designed to help readers enrich their learning experience. At the close of each chapter, the authors have developed a list of key concepts and terms taken from the chapter and two or three discussion questions to test comprehension and inspire classroom engagement. Instructors who adopt this book will also find the Internet sources, additional reading, and video selections of interest for developing meaningful assignments that allow students to explore these issues more thoroughly.

As a whole, the chapters in this volume provide new insights into the ways that actors in various regions around the world are reshaping their agrifood systems to realize the goal of sustainability. We hope that these case studies will demystify the process of agrifood change and stimulate an interest in carrying on this important dialogue, as well as reach out to new readers unfamiliar with these issues, for such readers are critical if the transformative promise of food system renewal is to be realized.

REFERENCES

Altieri, Miguel A. 2004. *Genetic Engineering in Agriculture: The Myths, Environmental Risks, and Alternatives.* Oakland, Calif.: Food First.

Andreatta, Susan, and William Wickliffe II. 2002. "Managing Farmer and Consumer Expectations: A Study of North Carolina Farmers Market." *Human Organization* 61 (2): 167–76.

Arce, Alberto, and Terry Marsden. 1993. "The Social Construction of International Food: A New Research Agenda." *Economic Geography* 69 (3): 293–311.

Archer, Margaret. 1982. "Morphogenesis Versus Structuration: On Combining Structure and Action." *British Journal of Sociology* 33 (4): 455–83.

Bangor Daily News. 2000. "Monsanto Lab in Crystal Closes Amid Food Protests." 3 May. www.biotech-info.net/monsanto_lab.html.

Bonanno, Alessandro, and Douglas H. Constance. 2000. "Mega Hog Farms in the Texas Panhandle Region: Corporate Actions and Local Resistance." *Research in Movements, Conflict, and Change* 22: 83–110.

Booth, David. 1994. "Rethinking Social Development: An Overview." In *Rethinking Social Development: Theory, Research, and Practice,* ed. David Booth, 3–34. Essex, UK: Longman Scientific and Technical.

Bourdieu, Pierre. 1984. *Distinction: A Social Critique of the Judgment of Taste.* Cambridge: Harvard University Press.

Brooks, David. 2001. *Bobos in Paradise: The New Upper Class and How They Got There.* New York: Simon and Schuster.

Brown, Allison. 2001. "Counting Farmers Markets." *Geographical Review* 91 (4): 655–74.

Brown, William P. 1988. *Private Interests, Public Policy, and American Agriculture.* Lawrence: University Press of Kansas.

Brubaker, Elizabeth. 2004. "As Property Rights Slide, Odours Rise." *Financial Post* (Canada), 23 April. www.hogwatchmanitoba.org/news0404.html.

Burns, Tom R., and Helana Flam. 1986. *The Shaping of Social Organization: Social Rule System Theory with Applications.* Beverly Hills, Calif.: Sage Publications.

Busch, Lawrence. 1980. "Structure and Negotiation in the Agricultural Sciences." *Rural Sociology* 45 (1): 26–48.

Busch, Lawrence, William B. Lacy, Jeffrey Burkhardt, and Laura R. Lacy. 1991. *Plants, Power, and Profit: Social, Economic, and Ethical Consequences of the New Biotechnologies.* Cambridge, Mass.: Basil Blackwell.

Buttel, Frederick H. 2000. "The Recombinant BGH Controversy in the United States: Toward a New Consumption Politics of Food." *Agriculture and Human Values* 17 (1): 5–20.

Carson, Rachel. 1962. *Silent Spring.* Boston: Houghton Mifflin.

Cochrane, Willard W. 1979. *The Development of American Agriculture: A Historical Analysis.* Minneapolis: University of Minnesota Press.

Cole, Wendy. 2007. "Got Raw Milk? Be Very Quiet." *Time* magazine, 13 March. www.time.com/time/health/article/0,8599,1598525,00.html.

Craib, Ian. 1992. *Anthony Giddens.* London: Routledge.

Critser, Greg. 2003. *Fat Land: How Americans Became the Fattest People in the World.* Boston: Houghton Mifflin.

Davidson, Osha Gray. 1996. *Broken Heartland: The Rise of America's Rural Ghetto.* Iowa City: University of Iowa Press.

DeLind, Laura B. 1998. "Parma: A Story of Hog Hotels and Local Resistance." In *Pigs, Profits, and Rural Communities,* ed. Kendall M. Thu and E. Paul Durrenberger, 23–38. Albany: State University of New York Press.

Dudley, Kathryn Marie. 2000. *Debt and Dispossession: Farm Loss in America's Heartland.* Chicago: University of Chicago Press.

DuPuis, E. Melanie. 2000. "Not in My Body: rBGH and the Rise of Organic Milk." *Agriculture and Human Values* 17 (3): 285–95.

Ehrenreich, Barbara. 1985. "Food Worship." In *The Worst Years of Our Lives: Irreverent Notes from a Decade of Greed,* ed. Barbara Ehrenreich, 18–20. New York: Pantheon Books.

Feenstra, Gail W. 1997. "Local Food Systems and Sustainable Communities." *American Journal of Alternative Agriculture* 12 (1): 28–36.

Fine, Ben. 1995. "From Political Economy to Consumption." In *Acknowledging Consumption: A Review of New Studies,* ed. Daniel Miller, 127–63. London: Routledge.

Fine, Ben, and Ellen Leopold. 1993. *The World of Consumption.* London: Routledge.

Food and Agriculture Organization of the United Nations (FAO). 2006. "The State of Food Insecurity in the World: Eradicating World Hunger—Taking Stock Ten Years After the World Food Summit." www.fao.org/docrep/009/a0750e/a0750e00.htm.

Friedland, William. 1994. "The New Globalization: The Case of Fresh Produce." In *From Columbus to ConAgra: The Globalization of Agriculture and Food,* ed. Alessandro Bonanno, Lawrence Busch, William H. Friedland, Lourdes Gouveia, and Enzo Mingione, 210–31. Lawrence: University Press of Kansas.

Giddens, Anthony. 1984. *The Constitution of Society: Outline of the Theory of Structuration.* Berkeley and Los Angeles: University of California Press.

Goodman, David. 1999. "Agro-Food Studies in the 'Age of Ecology': Nature, Corporeality, Bio-Politics." *Sociologia Ruralis* 39 (1): 17–38.

———. 2004. "Rural Europe Redux: Reflections on Alternative Agro-Food Networks and Paradigm Change." *Sociologia Ruralis* 44 (1): 3–16.

Goodman, David, and E. Melanie DuPuis. 2002. "Knowing Food and Growing Food: Beyond the Production-Consumption Debate in the Sociology of Agriculture." *Sociologia Ruralis* 42 (1): 5–22.

Goodman, David, and Michael Redclift. 1991. *Refashioning Nature: Food, Ecology, and Culture.* London: Routledge.

Goodman, David, Bernardo Sorj, and John Wilkinson. 1987. *From Farming to Biotechnology: A Theory of Agro-Industrial Development.* Oxford: Basil Blackwell.

Greene, Catherine, and Carolyn Dimitri. 2003. *Organic Agriculture: Gaining Ground.* Washington, D.C.: U.S. Department of Agriculture.

Groh, Trauger, and Steven McFadden. 1997. *Farms of Tomorrow Revisited: Community Supported Farms—Farm Supported Communities.* Kimberton, Pa.: BioDynamic Farming and Gardening Association.

Guthman, Julie. 2003. "Eating Risk: The Politics of Labeling Genetically Engineered Foods." In *Engineering Trouble: Biotechnology and Its Discontents,* ed. Rachel A. Schurman and Dennis Doyle Takahashi Kelso, 130–51. Berkeley and Los Angeles: University of California Press.

———. 2004. *Agrarian Dreams: The Paradox of Organic Farming in California.* Berkeley and Los Angeles: University of California Press.

Habermas, Jurgen. 1987. *The Theory of Communicative Action.* Vol. 2, *Lifeworld and System: A Critique of Functionalist Reason.* Boston: Beacon Press.

Hall, Jeff, Will Snell, and Craig Infanger. 2000. "Kentucky's (Phase I) Tobacco Settlement Funding Program." http://kytobaccotrust.state.ky.us/phase_ii/index.shtml.

Hayes, Matthew, and Kinga Milánkovics. 2001. *Community Supported Agriculture (CSA): A Farmers' Manual.* Gödöll?, Hungary: Szent István University, Institute for Environmental Management and Nyitott Kert Alapitvany.

Henderson, Elizabeth. 2000. "Rebuilding Local Food Systems from the Grassroots Up." In *Hungry for Profit: The Agri-business Threat to Farmers, Food, and the Environment,* ed. Fred Magdoff, John Bellamy Foster, and Frederick H. Buttel, 175–88. New York: Monthly Review Press.

Hendrickson, John. 1999. "Community Supported Agriculture: Growing Food . . . and Community." Research Brief No. 21. Madison: University of Wisconsin, Center for Integrated Agricultural Systems.

Hinrichs, C. Clare. 2000. "Embeddedness and Local Food Systems: Notes on Two Types of Direct Agricultural Market." *Journal of Rural Studies* 16 (3): 295–303.

Hinrichs, C. Clare, Gilbert Gillespie, and Gail W. Feenstra. 2004. "Social Learning and Innovation at Retail Farmers' Markets." *Rural Sociology* 69 (1): 31–58.

International Federation of Organic Agriculture Movements (IFOAM). 2007. "The World of Organic Agriculture: Statistics and Emerging Trends, 2007." www.ifoam.org/press/press/statistics2007.html.

Jackson, Laura. 1998. "Large Scale Swine Production and Water Quality." In *Pigs, Profits, and Rural Communities,* ed. Kendall M. Thu and E. Paul Durrenberger, 103–19. Albany: State University of New York Press.

James, Clive. 1998. "Global Review of Commercialized Transgenic Crops: 1998." ISAAA Briefs No. 8. Ithaca, N.Y.: International Service for the Acquisition of Agri-Biotech Applications.

Juska, Arunas, and Bob Edwards. 2004. "Refusing the Trojan Pig: The Trans-Atlantic Coalition Against Corporate Pork Production in Poland." In *Coalitions Across Borders: Transnational Protest and the Neo-Liberal Order,* ed. Joe Brandy and Jackie G. Smith, 187–207. Lanham, Md.: Rowman and Littlefield.

Kauffman, L. A. 1991. ""New Age Meets New Right: Tofu Politics in Berkeley." *The Nation,* 16 September, 294–96.

Kenney, Martin. 1986. *Biotechnology: The University-Industrial Complex.* New Haven: Yale University Press.

Kleiner, Anna M., and Douglas H. Constance. 1998. "Circling the Wagons: The Westward Expansion of Pork Production in Support of Circle Four." Paper presented at the annual meeting of the Rural Sociology Society, Portland, Oregon, August.

Kloppenburg, Jack R., Jr. 1988. *First the Seed: The Political Economy of Plant Biotechnology, 1492–2000.* Cambridge: Cambridge University Press.

Krimsky, Sheldon, and Roger P. Wrubel. 1996. *Agricultural Biotechnology and the Environment: Science, Policy, and Social Issues.* Urbana: University of Illinois Press.

Latour, Bruno. 1987. *Science in Action: How to Follow Scientists and Engineers Through Society.* Cambridge: Harvard University Press.

Lockie, Stewart, Kristin Lyons, Geoffrey Lawrence, and Kerry Mummery. 2002. "Eating 'Green': Motivations Behind Organic Food Consumption in Australia." *Sociologia Ruralis* 42 (1): 20–37.

Lyson, Thomas A. 2004. *Civic Agriculture: Reconnecting Farm, Food, and Community.* Medford, Mass.: Tufts University Press.

Lyson, Thomas A., Gilbert Gillespie, and D. Hilchey. 1995. "Farmers Markets and the Local Community: Bridging the Formal and Informal Economy." *American Journal of Alternative Agriculture* 10 (3): 108–13.

Marsden, Terry, and Alberto Arce. 1995. "Constructing Quality: Emerging Food Networks in the Rural Transition." *Environment and Planning A* 27 (8): 1261–79.

Marx, Karl. 1869/1963. *The Eighteenth Brumaire of Louis Bonaparte.* New York: International Publishers.

Miele, Mara. 2001. *Creating Sustainability: The Social Construction of the Market for Organic Products.* Wageningen, The Netherlands: Circle for Rural European Studies, Wageningen University.

Mikkelsen, Thomas R., Bente Andersen, and Rikke Bagger Jørgensen. 1996. "The Risk of Crop Transgene Spread." *Nature* 380 (6569): 31.

Mooney, Patrick H., and Theo J. Majka. 1995. *Farmers' and Farm Workers' Movements: Social Protest in American Agriculture.* New York: Twayne Publishers.

Murdoch, Jonathan, Terry Marsden, and Jo Banks. 2000. "Quality, Nature and Embeddedness." *Economic Geography* 76 (2): 107–25.

Murdoch, Jonathan, and Mara Miele. 1999. "'Back to Nature': Changing 'Worlds of Production' in the Food Sector." *Sociologia Ruralis* 39 (4): 465–83.

Nestle, Marion. 2002. *Food Politics: How the Food Industry Influences Nutrition and Health.* Berkeley and Los Angeles: University of California Press.

New Internationalist. 2002. "The Slow Food Movement: Slow Food Protests over Fast Foods and Food Safety." 1 March. www.newint.org/issue343/action.htm.

Newman, Maria. 2005. "Tests Confirm 2nd Case of Mad Cow Disease in the U.S." *New York Times,* 24 June.

Nord, Mark, Margaret Andrews, and Steven Carlson. 2004. "Household Food Security in the United States." Food Assistance and Nutrition Report (FANRR42), October. Washington, D.C.: U.S. Department of Agriculture, Economic Research Service.

Organic Monitor. 2006. "The Global Market for Organic Food and Drink: Business Opportunities and Future Outlook." www.organicmonitor.com/.

Organic Trade Association (OTA). 1999. "Fact Sheets." www.ota.com/.

Perkins, Jerry. 2004. "Kinder, Gentler, Food: Restaurants, Groceries, and Activists Force Farmers to Change Treatment of Their Animals." *Des Moines Register,* 7 March, 1A.

Raynolds, Laura T. 2002. "Consumer/Producer Links in Fair Trade Coffee Networks." *Sociologia Ruralis* 42 (4): 404–24.

Rissler, Jane, and Margaret G. Mellon. 1996. *The Ecological Risks of Engineered Crops.* Cambridge: MIT Press.

Ritzer, George. 1996. *Sociological Theory.* New York: McGraw Hill.

Scott, James C. 1985. *Weapons of the Weak: Everyday Forms of Peasant Resistance.* New Haven: Yale University Press.

Slign, Michael, and Carolyn Christman. 2003. "Who Owns Organic? The Global Status, Prospects, and Challenges of a Changing Organic Market." Rural Advancement Foundation International–USA, Pittsboro, North Carolina. www.rafiusa.org/pubs/OrganicReport.pdf.

Souza Silva, José de. 1994. "Agricultural Biotechnology in Developing Nations: Place, Role, and Contradictions." Paper presented at the annual meeting of the National Agricultural Biotechnology Council, East Lansing, Michigan, 23–24 May.

Strange, Marty. 1988. *Family Farming: A New Economic Vision.* Lincoln: University of Nebraska Press.

Thompson, Paul B. 1996. "Food Labels and the Ethics of Consent." *Choices* (1st quarter): 11–13.

———. 1997. *Food Biotechnology in Ethical Perspective.* London: Chapman and Hall.

Thu, Kendall M., and E. Paul Durrenberger, eds. 1998. *Pigs, Profits, and Rural Communities.* Albany: State University of New York Press.

U.S. Department of Agriculture (USDA). 2006. "Farmers Market Growth." www.ams .usda.gov/farmersmarkets/FarmersMarketGrowth.htm.

Van der Ploeg, Jan, and Henk Renting. 2004. "Behind the 'Redux': A Rejoinder to David Goodman." *Sociologia Ruralis* 44 (2): 233–42.

Van Esterik, Penny. 2005. "No Free Lunch." *Agriculture and Human Values* 22 (2): 207–8.

Waterman, Stanley. 1998. "Carnivals for Elites? The Cultural Politics of Arts Festivals." *Progress in Human Geography* 22 (1): 54–74.

Whatmore, Sarah, Pierre Stassart, and Henk Renting. 2003. "What's Alternative About Alternative Food Networks?" *Environment and Planning A* 35 (3): 389–91.

Whatmore, Sarah, and Lorraine Thorne. 1997. "Nourishing Networks: Alternative Geographies of Food." In *Globalising Food: Agrarian Questions and Global Restructuring,* ed. David Goodman and Michael J. Watts, 287–304. London: Routledge.

Wright, D. Wynne. 2004. "The Irrationality of Rational Hogs." *Great Plains Sociologist* 16 (1).

PART I

CONCEPTUAL FRAMEWORK

1

AGENCY AND RESISTANCE IN THE SOCIOLOGY OF AGRICULTURE AND FOOD

Alessandro Bonanno and Douglas H. Constance

While agency has occupied a prominent position since the classical period of sociology (i.e, Marx and Engels 1846/1990; Weber 1903/1947; Durkheim 1893/1964 and 1895/1982), more recent debates have forcefully reintroduced the topic (i.e., Archer 1995; Bourdieu 1998; Giddens 1990, 1994; Joas 2004). Often motivated by the charge that modern theory has been exhausted, contemporary sociologists have sought novel explanations about the capacity of humans to act and the opposition that their actions encounter. Because of the transcendental nature of these debates, efforts have been largely concerned with the relationship between agency and structure. Evolving from the debates in the 1960s and 1970s about the macro–micro connection, the probing of the "agency-structure integration" issue has engendered an abundance of contributions. These include famous contributions such as Anthony Giddens's (1984) theory of "structuration," which postulates that agency constructs structures but that structures condition agency in terms that both hamper but also enable agents; the late Pierre Bourdieu's (1977) dialectical relationship between agency and structure centered around the concepts of "habitus" and "field"; and Margaret Archer's (1995) approach, which contrasts with Giddens's in prescribing an analytical separation between agency and structure.

While often stimulating, these many contributions have also triggered intellectual "boredom." Some sociologists have interpreted this long engagement as an endless endeavor that needs to be reconsidered. In this respect, Immanuel Wallerstein, former president of the International Sociological

Department of Sociology, P.O. Box 2446, Sam Houston State University, Huntsville, Texas, 77341-2446. E-mail: soc_aab@shsu.edu or soc_dhc@shsh.edu.

Association, writes, "I believe that the antinomy of agency and structure is one more antinomy that leads to endless debates that I suspect are largely futile" (Wallerstein 2004, 317).

Departing from these abstract postures, sociology of agriculture and food debates approached agency pragmatically, as their participants' intellectual work was aimed at the construction and application of social justice and equality-centered forms of social arrangements. In their work the notion of agency has been often—albeit not exclusively—associated with resistance to dominant groups and established forms of socioeconomic development. This chapter documents this tradition through a brief review of the intellectual history of the sociology of agriculture and food, undertaken in the first two sections. The third section briefly discusses the theoretical contributions of the essays in this volume by William Friedland and Norman Long in the context of the intellectual history of the sociology of agriculture and food. The final section illustrates salient themes stemming from substantive research in agriculture and food. We conclude that the new patterns of food consumption and the shrinking regulatory role of the state can be fundamental forces in the shaping of new "fights over food."

The Sociology of Agriculture and Food

The pragmatic approach to the issue of agency of those who have participated in sociology of agriculture and food debates is largely the result of their long-standing concern for the search for more viable forms of "socioeconomic development." This search has focused on change, social justice, and the establishment of more equitable socioeconomic conditions, along with a direct engagement with the issue of resistance. In this context, agency has been understood as an essential component of the processes of resistance to, and transformation of, established socioeconomic arrangements.

INTELLECTUAL AND INSTITUTIONAL ORIGINS

To illustrate the evolution of the sociology of agriculture and food, we focus on the intellectual activities of the International Sociological Association's Research Committee on Sociology of Agriculture and Food, or RC-40.[1] To

---

1. RC stands for Research Committee and the number 40 indicates that this RC is the fortieth to join the International Sociological Association (ISA). Currently, the ISA has fifty-three RCs.

be sure, the development of the substantive area of the sociology of agriculture and food has been influenced by debates within and between a number of disciplines and intellectual circles. As such, the sociology of agriculture and food is clearly larger than the works produced by the members of RC-40. However, the quantity and quality of the contributions generated by this group allow us to employ RC-40 as an empirical indicator of the evolution of the debate within this substantive area. In addition, the creation of this institution provided a forum that linked regional and continental debates and fostered intellectual interaction across linguistic and cultural-historical milieus.

The sociology of agriculture became an openly discussed substantive area in the 1970s. The scientific topics that formed the group's agenda, however, were not new, as they have their roots in debates centered on classical contributions, including those of Marx, Weber, Durkheim, and Kautsky. What was clearly new was the focus employed by these researchers. In much of the developed and developing world, debates around social issues concerning agriculture became increasingly severed from the established research foci of "rural" and "community." Emphasis shifted to issues such as social change, labor conditions, production structures in agriculture, and food availability and quality. These new interests evolved in a context in which dominant views of society became increasingly the subjects of intense scrutiny. In a number of national debates, modernization theory and its prescription for the *Americanization* of society were strongly criticized. Many of these critiques were fueled by emerging neo-Marxian (i.e., dependency, world-system, and critical theory) and constructionist (i.e., symbolic interactionism and phenomenology) theories. In both cases, the epistemological and normative components of the functionalist-based modernization theory were countered by historically grounded analyses that stressed the increasing socioeconomic gap between and within world regions and the problematic nature of power distribution in decision-making processes. Works based on these novel approaches increasingly appeared in regional, national, and, eventually, international conferences and disciplinary journals through the scientific activity of a group of young and well-prepared sociologists.

Those who constituted the North American group—and later those who provided impetus for the debate in Oceania (e.g., Lawrence 1987; McMichael 1984)—distanced themselves from the view that considered "rural society" as parallel to, and largely independent from, the rest of society. They took the position that agriculture and its subcomponent, farming, are parts of the "totality" of the evolution of capitalism. In this context, they viewed agriculture as experiencing concentration and centralization of power along with social marginalization (e.g., Friedland 1982). Adopting a similar framework, European and Latin American debates focused on the peasant question[2] and the transition to more advanced forms of capitalist agriculture (e.g., Chombart de Lauwe 1979; Chonchol 1994; Clout 1975). These debates were informed by critical views of the growth of capitalism and the search for solutions to the long-standing problem of underdevelopment of agriculture-based regions. While the different types of socioeconomic development in North America, Oceania, Europe, and Latin America[3] shaped the direction of the debate considerably on these continents, common features were the use of rich theoretical approaches and the reliance on critically based sociological theories. The socioeconomic background of these intellectual debates was the crisis of the postwar Fordist system and its political regime of Pax Americana, along with the emergence of patterns that later constituted the globalization project.

In the late 1970s, a limited number of international forums were amenable to the intellectual discussion fostered by members of the sociology of agriculture group. At the time, the International Rural Sociological Society (IRSA) was a nascent organization of regional rural sociological societies that did not recognize individual membership or allow the organization of research groups. Moreover, its activities were almost exclusively confined to the World Congress of Rural Sociology, which met every four years. Mostly because of this structure and its limited visibility between the world congresses, IRSA did not attract significant or organized participation of scholars working in the sociology of agriculture at the time. One of the other large sociological organizations, the International Sociological Association, had a much longer history and not only allowed individual memberships but fostered participation through the presentation of research grouped around topical themes. This open form of organization was key in the institutional development of the field of sociology of agriculture.

ORIGINAL THEORETICAL COMPONENTS

By the early 1980s, the critique of modernization theory was accompanied by an equally strong scrutiny of the tenets of orthodox Marxism. In particular,

2. Simply put, the peasant question refers to the debate on the evolution of precapitalist agriculture (i.e., peasant agriculture) into capitalist agriculture (i.e., farmers who, through the control of labor and the means of production, produce commodities for the market). In particular, the peasant question dealt with the obstacles that prevented peasant communities and relations of production from evolving into capitalist (i.e., market) ones, or delayed that transition.

3. While contributions from Africa were and still remain largely absent from these debates, a number of important contributions from Asia entered the debate in the late 1980s and 1990s.

this critique was directed at the Leninists'"green factory" thesis of the development of agriculture.[4] According to this thesis, agriculture would evolve into a sector in which the relations of production would resemble those of industrial capitalism: large factories employing a large number of wage laborers. This capital-intensive and wage-based agriculture would be the "norm" in advanced societies and would gradually expand to the rest of the world.

In this context, an analysis of the contemporary forms and relevance of the "agrarian question" took center stage in the debate among the members of the sociology of agriculture group.[5] It entailed the use of novel and more sophisticated theoretical frameworks that would permit an intellectual distancing from the equally inadequate modernization theory and Marxist-Leninist formulations. To this end, a number of classical contributions were reintroduced and employed to analyze the agrarian question in last quarter of the twentieth century, among them those of Karl Kautsky, Aleksandr Chayanov, and Antonio Gramsci, along with alternative interpretations of Max Weber. In addition, the new sociology of agriculture—as it became known, to emphasize the changed focus of the debate—featured theorizations that found their roots in phenomenology, interactionism, and nascent postmodern pronouncements. These constructionist approaches stressed the negotiated and relativistic dimensions of agricultural production and consumption and viewed agency in terms of people's ability to construct "realities" and define them as objective components of their everyday existence. Because of their success, these readings of agriculture established themselves as a powerful intellectual complement to the neo-Marxian analyses more popular at the time.

Significantly, one of the central characteristics of both neo-Marxian and constructionist approaches was the emphasis on the role of agency in the development of, and resistance to, capitalist social relations. This group of *critical* sociologists approached the issue of agency largely in terms of the "peasant question" in the South and the issue of the structure of agriculture in the North. In the North, the theme of the crisis of family farms, but also

4. The Leninist position derived from Lenin's analysis of turn-of-the-century agriculture. Employing secondary data, Lenin demonstrated that the concentration of capital and the proletarianization of labor that were typical of industrial relations occurred in similar fashion in agriculture (see Lenin 1974). At the turn of the century and within Marxian circles, this posture was criticized by Karl Kautsky in his seminal book *Die Agrarfrage*. Kautsky rejected the thesis of the "green factory" and argued for the persistence of family farms. For Kautsky, the persistence of these farms was linked to both economic but also social trends, such as the ruling-class desire to control excessive unemployment and the contradictions of capitalism.

5. The agrarian question refers to issues associated with social relations stemming from the presence of people on the land. Inspired by Kautsky's work ("the agrarian question" is the English translation of *die Agrarfrage*), it centered on the trajectories of development in modern agriculture.

their persistence, occupied center stage. In this context, agency was relevant in at least two ways. First, and in micro terms, agency was the focal point of entry for the study of people on the land. It was stressed that farmers, laborers, and peasants do not necessarily make sense of the reality around them in the same way implied by dominant conceptualizations. Consequently, they do not automatically adopt rationales and behaviors that conform to dominant narratives. In fact, they often tend to reject dominant views and through these rejections establish practices of resistance.

Second, and in macro terms, agency was considered key in regard to processes of change. On the one hand, research questions gravitated around the possibility for the emergence of emancipatory action in a context that featured increased concentration of capital and political control by large transnational agrifood corporations. On the other hand, researchers focused on the ways in which farmers, laborers, and peasants actively responded to the restructuring of agriculture.

The Institutional Beginnings and Evolution of the Debate

The intellectual debate outlined above created the conditions for the development of RC-40. The first organizational step was taken at the Ninth World Congress of the International Sociological Association (ISA), held in Uppsala, Sweden, in July 1978. During the course of that conference, participants decided to create a permanent organizational structure within the ISA. The ISA recognized three types of research-based groups: ad hoc committees, working groups, and research committees (RCs). Each of these structures features progressively more advanced organizational forms that differ by quantity and quality of research activities, dissemination of knowledge, membership size, and number of officers. RCs are the most advanced structures within the ISA. The decision made in Uppsala resulted in the constitution of an ad hoc committee whose status was formally recognized in 1982 at the Tenth World Congress of Sociology, held in Mexico City. During that conference, the ad hoc committee on the sociology of agriculture organized twelve sections and elected Lawrence Busch to serve as the group's first president. Busch served for two consecutive four-year terms (1986–94), and under his leadership the group completed all the necessary requirements for elevation to the status of research committee. By early 1992 the sociology of agriculture was officially recognized as an ISA research committee. It is important to note that during this time the term "food" was added to the official name of the organization. This move responded to the acceleration of capital concentration in agriculture and the increased control exercised by large, and often transnational, agrifood corporations. The scientific point that accompanied the name change referred to the impossibility of meaningfully separating the study of agriculture production from that of food consumption.

Lawrence Busch's successor as president of RC-40, William Friedland, served for one term only (1994–98). Under his presidency, a number of conferences and miniconferences were organized around such themes as globalization and the role of the state in agriculture and food. In addition, the group published multiauthor books that provided impetus to the scientific debate. This intellectual climate fostered growth of the dialogue between neo-Marxian and constructionist approaches. Following the tradition of the group, this dialogue did not simply involve the generation of theoretical discussion but promoted postures that framed empirical research. Neo-Marxian approaches retained modern concerns for the development of analyses that could lead to the construction of alternative forms of development and address centralization of power, concentration of capital, and limited participation in decision-making processes. In particular, these analyses focused on the restructuring and crisis of the nation-state and the consequences of reduced state intervention in, and regulation of, society. Additionally, the emergence of the globalization project and its impact on food production and consumption occupied center stage in neo-Marxian works (e.g., Bonanno et al. 1994; Friedland et al. 1991; McMichael 1994).

Two types of approaches characterized constructionist contributions. The first was grounded in the phenomenological and interactionist traditions. Exemplified by the work of Busch and his associates (Busch and Lacy 1983, 1986; Busch et al. 1991), this approach proposed a critique of agricultural science—biotechnology in particular—and scientific institutions stressing the negotiated dimension of reality. The other approach, exemplified by the work of Norman Long and his associates (Long and Long 1992), found its roots in the postmodern actor-oriented focus. Emphasizing epistemological scrutiny of the limits of positivism and structuralism, it provided a strident critique of both modernization theory and some forms of neo-Marxism that opened the way to a wealth of pathbreaking contributions.

The dialogue between the neo-Marxian and constructionist camps continued through the 1990s and into the new century under the presidencies of Philip McMichael (1998–2002) and Manuel Belo Moreira (2002–6). These years featured a much more focused effort to address issues pertaining to the global liberalization of markets, the growth of transnational corporations and corporate retailers, the evolution of regulations and standards in agriculture and food, and resistance to dominant patterns of agricultural and food production, including the introduction of genetically modified foods and corporate control of food consumption.

Theoretical Contributions

The essays contained in this volume reflect the sociology of agriculture and food group's continued emphasis on the search for alternative social arrangements and social justice. At the level of theory, the contributions of Friedland and Long, which follow this chapter, are examples of substantive-oriented theory. These are instances of the group's abiding concern with the creation of theoretical frameworks that can be employed in the development of empirical investigations. Writing from a neo-Marxian perspective, Friedland argues that the renewed debate on the issue of agency is the product of changed socioeconomic conditions. These conditions made established theories inadequate, as the historical subject of agency and resistance—the classical proletariat—has exhausted its emancipatory role. In the Marxian paradigm, agency is that collective dimension that is necessary to generate resistance to capitalism's dominant classes. While agency's characteristics and modes of historical manifestation have varied, the fundamental tenet of Marxian theories has been that class membership and struggle determine the emergence of agency and, through agency, resistance to dominant groups. Friedland upholds this tenet.

It should be added that over the years the Marxian camp has featured a number of readings on agency and resistance. In the theories of economic determinism in vogue in the late nineteenth century, agency was seen as the immediate product of class polarization. The worsening of the socioeconomic conditions of the working class would almost automatically generate the necessary conditions for the development of anticapitalist revolutionary actions (e.g., Kautsky 1892/1971). The Leninist critique of this position viewed agency as the exclusive prerogative of the Communist Party vanguard. This group would lead the proletarian masses to resistance against capitalism and the establishment of socialism. The crisis of the Soviet model engendered harsh critiques of Leninism, in terms of both a participatory view of agency (i.e., agency in the acting masses [Luxemburg 1951]) and a culturally centered notion of agency (agency as the product of consciousness raising [e.g., Gramsci 1971; Lukács 1971]). In addition, the changed characteristics of capitalism in the second half of the twentieth century paved the way for the development of another round of—now neo—Marxian interpretations, including critical theory (agency as also the prerogative of the individual in a class society [e.g., Marcuse 1964]) and structuralism (agency as determined by the structure of capitalism [e.g., Althusser and Balibar 1970/1998]). All of this is to say that, in the absence of convincing Marxian theories, Friedland's contribution focuses on the intellectual task of identifying the historical conditions that allow for the manifestation of agency and resistance.

The same desire to provide theoretical assistance to the substantive study of agency and resistance motivates Norman Long's essay. In the last three decades, Long's actor-oriented approach has been employed to produce important sociological studies that have provided relevant theoretical insights into the study of agriculture and food. These intellectual contributions also focus on the "limits" of structuralist accounts and Marxian approaches.

The actor-oriented critique of structuralism is centered on the antifoundational posture maintained by this approach. Antifoundationalism signifies the rejection of the idea that action can be ultimately linked to any underlying structure. While several structuralist approaches exist—chiefly functionalist and semiotic—neo-Marxian works in agriculture and food have been more common and have tended to attribute to economic forces (i.e., the so-called "logic of capital") the ultimate capacity to shape action. This "rigidity" of structuralist approaches prevents the understanding of the relative position of the actor in regard to other elements that form the networks (contexts) within which action is performed. According to the actor-oriented approach, understanding action is not about understanding its essential dimension; rather, it is the process of learning from actors the features of their actions (what they are doing) and the motivations behind their actions (why they are doing it).

The exquisitely modern insistence of neo-Marxian approaches on searching for grand narratives has often been the source of interpretations that tend to simplify complex situations. While heuristic simplification is not per se to be condemned, the elimination of complexity and the accompanying silencing of alternative voices are problematic. To contrast this totalizing "narrowness" of Marxism, the actor-oriented approach proposes a reading of agency that acknowledges the diversity and complexity of positions from which actors approach and make sense of events around them. This construction transcends the simplistic "indigenous-based" reading of resistance that confines itself to the tenet that actors develop their own interpretations of reality by contesting dominant propositions. It mandates a much more relational reading of action in which alternative rationales and postures come to the fore through interfacing with dominant discourses and frames that ultimately create the common ground on which resistance takes place.

Change in Agriculture and Food

Let us turn now to the salient changes under way in the contemporary agri-food system. Our observations are not intended to be exhaustive, but we hope that they can be of assistance in reading the chapters that follow.

I. LARGE CORPORATIONS EXERCISE GREATER CONTROL AT THE LEVEL OF PRODUCTION

This is the result of globalization that features the implementation of neo-liberal policies that have opened markets, diminished state intervention and regulation, enhanced competition, and expanded the circulation of commodities and capital. The hypermobility of capital permitted by globalization—the compression of time and space that allows the acceleration of the reproduction of capital—allows transnational corporations to enhance their ability to search for more convenient conditions of production. This new organization of production has entailed the fragmentation of labor in terms of both its physical utilization—through decentralized production processes—and its political strength—the crisis of unions and working-class political parties and programs (Coates 2000; Western 1995). Simultaneously, the opening of global markets has required communities to adopt new strategies to remain economically viable. The most frequently employed are the discounting of local natural resources, infrastructures, and the price of social services, the introduction of tax abatements, and the adoption of general procorporate political measures, all financed largely by public resources. Because historically labor- and class-based movements have identified production processes as the primary site of exploitation and the arena in which to concentrate struggle, their fragmentation indicates that current conditions provide corporations with additional control of, and strength at, the production level.

2. CURRENT CONDITIONS WEAKEN RESISTANCE AT THE LEVEL OF PRODUCTION

Enhanced corporate control at the production level has resulted in the decreased ability of historical working-class movements, parties, and unions to mobilize resistance to dominant social groups. In the past, working-class organizations' most important gains came from their ability to organize and find strength in factories, agricultural fields, and other places of production. Once production was decentralized across distant areas and in dispersed subunits, labor pools were pitted against faraway counterparts, and workers' sense of solidarity and cooperation was severely diminished by downsizing and transnational competition, with the result that labor and its class-based ideology lost most of their power. While agrifood production remains a highly differentiated sector, the ability to resist dominant production arrangements is made more problematic by corporate hypermobility (Bonanno and Constance 2000).

3. CURRENT CONDITIONS ENHANCE THE IMPORTANCE OF CONSUMPTION

While past scholarship viewed consumption as a secondary component of the process of capital accumulation and social exploitation, more recent interpretations have focused on consumption as a vehicle of renewed domination, but also emancipation.

a. Consumption and Domination

Initiated by the critique of consumerism, the idea that subordinate masses are pacified through corporate-manipulated consumption is not a new proposition (see Marcuse 1964). In recent years, however, it has assumed new connotations stemming from the emergence of consumption patterns that are inspired by alternative political postures. In the case of food, consumption that emphasizes quality, naturalness, tradition, locality, and environmental stewardship has reached unprecedented proportions. This expansion has caught the attention of corporate retailers, which have adopted strategies to capture demand for high-quality food. Facilitated by the retreat of the regulatory state (see below), corporate retailers have enforced new standards that, while appeasing consumers, have also significantly affected producers. In particular, producers around the world who cannot meet these corporate requests have been excluded from production circuits (Marsden 2003). Simultaneously, critics have raised questions about the emancipatory dimension of consumption (see Jameson 1994; Eagleton 1996). The hegemonic powers of transnational corporations transform new and augmented consumption into novel markets that they use to expand profit.

b. Consumption and Emancipation

While this new consumption can be conceptualized in terms of the agency of individual consumers, it is the role played by new social movements that deserves attention. In the case of agriculture and food, the emergence of new social movement–based demands for environmentally sound, high-quality, healthier food items have affected the production, distribution, and social sites of consumption. Because of the large economic power associated with these types of consumers, transnational corporations have been forced to be attentive to their demands. In what is arguably an extreme rendering of this situation, Bauman (1992) maintains that consumption-substituted labor is the primary social site where identity and behaviors are formed. Humphery (1998) and Lang (1999) stress the emerging power of the "food movement," which, despite setbacks, "has been successful in questioning the New Right logic of laissez-faire and deregulating" (Lang 1999, 175; see also Gabriel and Lang 1995; Miller 1995). In this context, consumers can be viewed as much more thoughtful about the content and value of their consumption. This "reflexivity" about the conditions of consumption becomes an emancipatory process that contributes to the creation of free spaces in society. Josée Johnston pursues this question of consumption as counterhegemony in Chapter 4.

4. THE RETREAT OF THE STATE AS A REGULATOR: DEREGULATION OF THE PUBLIC SPHERE AND NEW REGULATION IN THE PRIVATE SPHERE

Owing to its crisis, the nation-state has become increasingly—albeit not totally—unable to act as a counterpart to the economic and social demands of subordinated groups. The state's decreased capacity to control transnational corporations, and a broad array of unwanted consequences of capitalism— e.g., unemployment, underemployment, social degradation, food quality— have sharply lessened the effectiveness of subordinate groups' claims against it. In this context, the state has become much less able to direct the socio-economic development of needy groups and regions. The management of projects such as community development, employment opportunities, and regional planning have become increasing difficult for agencies of the state, as they lack the instruments to establish sufficient control over the most relevant components of development. While state officialdom is actively disempowering the state, important prerogatives of the state are increasingly adjudicated in the private sphere, with the result that subordinate stakeholders are increasingly excluded from decision-making processes.

Conclusions

The current status of agriculture and food points to the emergence of new actors. On one side, transnational corporations, through their hypermobility,

are increasingly shaping the production and distribution of food items globally. On the other side, new forms of consumption have acquired importance.
This is the type of consumption that Lash and Urry (1994) called "reflexive
consumption" and Humphery (1998) described as empowering consumers
and pushing them toward cultural subversion and political resistance. It
mandates consumer behaviors that are inspired by new social movements'
sensitivities and therefore pay particular attention to the forms of production and the social contents of products. It questions key negative features
of global corporate production such as sweatshop-based manufacturing, the
use of child labor, the exploitation of women in *maquiladora* plants, the abuse
of human and environmental resources, and the destruction of cultural en-
claves; and it proposes better-quality, healthier, more culturally sensitive and
environmentally sound products. More important, this type of consumption
criticizes transnational corporations that favor quantity-based mass production over quality.

The essays in this volume reveal that new consumption-based social movements have experienced some success, yet scrutiny of these movements—
along with the actors involved—is essential. The extent to which these new
sensitivities translate into actual resistance, and the question of whether they
are simply new ways of enhancing corporate profit and control, should be
addressed seriously in the investigation of agriculture and food. Simultaneously, the retreat of the state and its consequences raise new questions that
must not be overlooked. The growing role of corporate actors in the shaping of the production and consumption of food is one of the emerging substantive areas that require scientific and social attention in the panorama of
our new fights over food.

REFERENCES

Althusser, Louis, and Etienne Balibar. 1970/1998. *Reading Capital.* London: Verso.
Archer, Margaret. 1995. *Realist Social Theory: The Morphogenetic Approach.* Cambridge: Cam-
        bridge University Press.
Bauman, Zygmunt. 1992. *Intimations of Postmodernity.* New York: Routledge.
Bonanno, Alessandro, Lawrence Busch, William H. Friedland, Lourdes Gouveia, and Enzo
        Mingione, eds. 1994. *From Columbus to ConAgra: The Globalization of Agriculture and
        Food.* Lawrence: University Press of Kansas.
Bonanno, Alessandro, and Douglas H. Constance. 2000. "Powers and Limits of Transna-
        tional Corporations: The Case of ADM." *Rural Sociology* 65 (3): 440–60.
Bourdieu, Pierre. 1977. *Outline of a Theory of Practice.* Cambridge: Harvard University Press.
———. 1998. *Practical Reason.* Stanford: Stanford University Press.

Busch, Lawrence, and William B. Lacy. 1983. *Science, Agriculture, and the Politics of Research.* Boulder, Colo.: Westview Press.

———. 1986. *The Agricultural Scientific Enterprise.* Boulder, Colo.: Westview Press.

Busch, Lawrence, William B. Lacy, Jeffrey Burkhardt, and Laura R. Lacy. 1991. *Plants, Power, and Profit: Social, Economic, and Ethical Consequences of the New Biotechnologies.* Cambridge, Mass.: Basil Blackwell.

Chombart de Lauwe, Louis. 1979. *L'Aventure Agricole de la France de 1945 a nos Jours.* Paris: PUF.

Chonchol, Jacques. 1994. *Sistemas Agrarios en America Latina.* Mexico City: Fondo de Cultura Economica.

Clout, David. 1975. *Regional Development in Western Europe.* London: John Wiley & Son.

Coates, David. 2000. *Models of Capitalism: Growth and Stagnation in the Modern Era.* Cambridge, UK: Polity Press.

Durkheim, Emile. 1893/1964. *The Division of Labor in Society.* New York: Free Press.

———. 1895/1982. *The Rules of Sociological Method.* New York: Free Press.

Eagleton, Terry. 1996. *The Illusion of Postmodernism.* Cambridge, Mass.: Blackwell.

Friedland, William H. 1982. "The End of Rural Society and the Future of Rural Sociology." *Rural Sociology* 47 (4): 589–608.

Friedland, William H., Lawrence Busch, Frederick H. Buttel, and Alan Rudy. 1991. *Towards a New Political Economy of Agriculture.* Boulder, Colo.: Westview Press.

Gabriel, Yiannis, and Tim Lang. 1995. *The Unmanageable Consumer.* London: Sage Publications.

Giddens, Anthony. 1984. *The Constitution of Society: Outline of the Theory of Structuration.* Berkeley and Los Angeles: University of California Press.

———. 1990. *The Consequences of Modernity.* Stanford: Stanford University Press.

———. 1994. *Beyond Left and Right.* Stanford: Stanford University Press.

Gramsci, Antonio. 1971. *Selections from the Prison Notebooks.* Ed. and trans. Quintin Hoare and Geoffrey Nowell Smith. New York: International Publishers.

Humphery, Kim. 1998. *Shelf Life: Supermarkets and the Changing Cultures of Consumption.* Cambridge: Cambridge University Press.

Jameson, Fredric. 1994. *Postmodernism or the Cultural Logic of Late Capitalism.* Durham: Duke University Press.

Joas, Hans. 2004. "The Changing Role of the Social Sciences: An Action-Theoretical Perspective." *International Sociology* 19 (3): 301–14.

Kautsky, Karl. 1892/1971. *The Class Struggle.* New York: W. W. Norton.

Lang, Tim. 1999. "The Complexity of Globalization: The UK as a Case Study of Tensions Within the Food System and the Challenge to Food Policy." *Agriculture and Human Values* 16 (2): 169–85.

Lash, Scott, and John Urry. 1994. *Economies of Signs and Space.* London: Sage Publications.

Lawrence, Geoffrey. 1987. *Capitalism and the Countryside.* Sydney: Pluto Press.

Lenin, Vladimir Ilyich. 1974. *The Development of Capitalism in Russia.* Moscow: Progress Publisher.

Long, Norman, and Ann Long, eds. 1992. *Battlefields of Knowledge: The Interlocking of Research and Practice in Social Development.* London: Routledge.

Lukács, Georg. 1971. *History and Class Consciousness.* Cambridge: MIT Press.

Luxemburg, Rosa. 1951. *The Accumulation of Capital.* New York: Routledge.

Marcuse, Herbert. 1964. *One-Dimensional Man.* Boston: Beacon Press.

Marsden, Terry. 2003. *The Condition of Rural Sustainability.* Assen, The Netherlands: Royal Van Gorcum.

Marx, Karl, and Friedrich Engels. 1846/1990. *The German Ideology.* New York: International Publishers.

McMichael, Philip. 1984. *Settlers and the Agrarian Question.* New York: Cambridge University Press.

———, ed. 1994. *The Global Restructuring of Agro-Food Systems.* Ithaca: Cornell University Press.

Wallerstein, Immanuel. 2004. "The Actor in the Social Sciences: A Reply to Hans Joas." *International Sociology* 19 (3): 315–21.

Weber, Max. 1903/1947. *The Methodology of the Social Sciences.* New York: Free Press.

Western, Bruce. 1995. "A Comparative study of Working-Class Disorganization: Union Decline in Eighteen Advanced Capitalist Countries." *American Sociological Review* 60 (2): 179–201.

INSTRUCTOR'S RESOURCES

Key Concepts and Terms:
1. Sociology of Agriculture and Food
2. Theories of agency
3. Globalization
4. Consumption

Discussion Questions:
1. Illustrate the origins, tenets, and theoretical developments of the debate in the sociology of agriculture and food.
2. Identify and discuss recent changes in agriculture and food.
3. Illustrate the basic tenets of the neo-Marxian and constructionist camps.

Agriculture, Food, and Environment on the Internet:
1. International Sociological Association Research Committee on Sociology of Agriculture and Food (RC-40): www.ucm.es/info/isa/rc40.htm.
2. International Journal of Sociology of Agriculture and Food: www.csafe.org.nz/ijsaf/.
3. International Rural Sociological Association (IRSA): www.irsa-world.org/.

Additional Readings:
1. Antonio, Robert, and Alessandro Bonanno. 2000. "A New Global Capitalism? From 'Americanism and Fordism' to 'Americanization-Globalization.'" *American Studies* 41 (2–3): 33–77.
2. Bonanno, Alessandro, and Douglas H. Constance. 1996. *Caught in the Net: The Global Tuna Industry, Environmentalism, and the State.* Lawrence: University Press of Kansas.
3. Friedland, William H., Lawrence Busch, Frederick H. Buttel, and Alan Rudy, eds. 1991. *Towards a New Political Economy of Agriculture.* Boulder, Colo.: Westview Press.
4. Long, Norman, and Ann Long, eds. 1992. *Battlefields of Knowledge: The Interlocking of Theory and Practice in Social Research and Development.* London: Routledge.

2

AGENCY AND THE AGRIFOOD SYSTEM

William H. Friedland

We are so deeply embedded in cultural traditions that, even as we transform and attack them, we cannot escape them. . . . For transform them we do. . . . Culture is only one dimension of life that humans are capable of changing. . . . This artful creativity—sociologists refer to it as agency—*is present in all social life, not just in protest.*

—JASPER (1997, 11)

Two tasks are undertaken in this chapter. First, I explore the current interest in agency and speculate as to its emergence, particularly among social scientists involved in the production-consumption debates. Second, because of the lack of definition of agency, I give various empirical examples in which behaviors related to agency can be said to be at work. These cases are used to develop a typology for empirical research.

In everyday life, people make hundreds of decisions. Most are made without deep reflexive behavior. We get up in the morning, do our toilette, make or buy our breakfast, eat it with little thought. Surely this is not agency but a reflection of structure: culture, economics, personal history or habit, etc. The distinction between structure and agency has plagued philosophers for several millennia, usually about religious questions. Currently, agency has become an issue for social scientists and humanists in nonreligious contexts. This chapter confronts the structure/agency issue in the agrifood sector. By focusing on agriculture and food, the basics of everyday existence, we may find something useful in understanding agency.

As debate about agency has recently unfolded, what has been striking has been the vagueness of conceptualization, the weight of philosophical debate, and the lack of empirical analysis.[1] I have been reluctant to begin research on

---

1. While the use of the term *agency* has proliferated, a paper by my colleagues Margaret Fitz-Simmons and David Goodman (1998) prompted a substantive reaction. The authors, emphasizing

how agency arose anew in social science, literary, and critical theory circles. Part of this is a personal disinclination to get lost in the history of ideas; I am more comfortable with research on empirical matters. I prefer to get at questions such as: what is the importance of agency to us now, in our world? How might we think about agency; particularly, how might it be useful in furthering the struggle for liberation, social justice, and equality?

This chapter consists of two parts: first, a review of the concept of agency, its origins, its demise—or, perhaps more accurately, its being made equivalent to the organized political party—and its resuscitation in the mid-1980s, plus some speculation as to the reasons for its phoenix-like return. Second, by using several cases, an exploration of an empirically based typology of agency. The empirical cases suggest:

1. Agency always has some anti- or counterhegemonic content. Action based on acceptance of the established hegemony, whether voluntary, induced, or coerced to accepted hegemonic norms, reflects structure, not agency.

2. Several continua can be found within the concept of agency. These include individual-group/social; levels or degrees of counterhegemony (CH), from innocuous to actively opposing hegemony, which also implies a continuum of resistance; and degrees of spontaneity, from totally spontaneous to highly organized.

3. There are at least two basic types of agency: progressive and reactionary. There are considerable numbers of individuals and groups that reject established hegemony and engage in counterhegemonic behavior that can be characterized as reactionary. For example, at the moment, the notion of intelligent design is counterhegemonic because it rejects the scientific hegemony of the concept of evolution. Since there is no scientific basis for the idea of intelligent design, I consider it reactionary; it throws belief back on faith. Similarly, there are movements that hearken back to an ostensible past and would seek to "re-create" that past. For my purposes, this chapter focuses on progressive agency,

---

the importance of "incorporating nature" into human affairs rather than thinking of nature as something separate from humans, went overboard, in my view, by attributing agency to nature and referring to "nature's agency": "Nonhuman nature is an actor," and "nature as an independent actor." There is a difference between nature being an actor in human affairs—causing humans to cope with the effects of a tornado or earthquake—but this hardly gives agency to nature. In my view, agency is a human attribute, the product of reflexive consciousness and having some counterhegemonic content.

although I acknowledge that much of the analysis could be revised to fit reactionary agency.

A variety of manifestations of progressive agency will be explored empirically, from the individual to the social and from those that have few or no consequences for social change to those that are profound for change.

Birth, Death (Co-optation), and Rebirth of Agency

The concept of agency had antecedents long before Marx. Theologians debated, centered around issues of free will, the choices individuals made to live by the word of God or to be sinful. While there are undoubtedly relationships to the present debate, I will forego any attempt to encompass them. For me, agency begins with Hegel and Marx. Hegel, because he built his philosophic analysis around the dialectic (any situation gives rise to its own contradiction); Marx, for linking the dialectic to the conception of class struggle.

I will not recapitulate how Europe's burghers, in establishing themselves as a bourgeoisie, created the proletariat, or how colonial administrators, seeking to reduce the costs of administration, created categories such as "African" and "Indian" and laid the basis for a nationalism that would be embodied in a cheaper educated native labor force. That each situation created contradictions was fundamental to Marx.

Marx understood the dialectic of the working class: that, as part of its class struggle, workers would create a complex of institutions to deal with everyday problems of life. These would include unions to confront employers; cooperatives to handle consumption problems; a paraphernalia of organizations to satisfy the needs and hopes of people for a cultural life (hiking organizations, singing societies, organizations to satisfy cultural interests in stamps, coins, the study of art or music); occupational organizations of lawyers, doctors, and other professionals, and so on. Capping this complex of organizations would be the political party of the working class. If agency bred the various proletarian organizations for specific purposes, the party would become the political agent of the class. Marx regarded these organizational formations as expressions of working-class agency.

What started as a philosophical spin-off of the dialectic became an embedded complex of organizations, so that by the time I personally experienced the socialist movement, the word "agency" had disappeared from discourse. We were so busy working in the specific agent intended to overthrow capitalism, the party, that no one was concerned with the meaning of agency. Individuals in the working class expressed their agency by joining the organized expressions of the class. "Agency" as such was never discussed; it had simply disappeared from leftist discourse.[2] Nor was agency a matter of concern to the New Left, since it had its organized expression in Students for a Democratic Society (SDS), which became, in effect, the "party" of the late 1960s and early 1970s.

The eruption of debate about agency in the 1980s seems to have been generated by Anthony Giddens (1984), who took up, as a sociologist with leftist inclinations, where Talcott Parsons (1951) had left off with his theory of action. Its recent revival comes from individuals and groups of scholars who have maintained a relationship to Marxism and claim to be the inheritors of the Marxist canon. This has been all the more puzzling since the party (and the auxiliary organizational complex associated with it) had effectively disappeared as an agent of change. Moreover, union membership—the economic manifestation of the proletariat that we had believed to be a fundamental driving force for change—was declining, and industrial workers in recent decades had never manifested the revolutionary urges that Marxists believed they were supposed to embody.

Agency and the Marxist Experience

Although a somewhat mystical concept that, from time to time, enthralls theologians and theorists of a Marxist persuasion, agency is not a clearly defined concept. While agency may be present in everyday life, it only infrequently rises to a level where it becomes publicly and broadly recognized.

There are situations in which agency seems to appear but at the same time circumscribe many of the instances that agency aficionados rely upon to sustain their beliefs that agency is always operative and becoming important or dominant. The intent of this chapter is to develop an empirically derived conception of agency and distinguish it from purely individual behaviors or more concrete and important social manifestations, social movements.

2. Agency has not been an explicit analytic category in Marxist analysis, although the term is implicit in much of Marxism. A search of the indexes of all forty-nine volumes of the *Collected Works* of Marx and Engels (1975–2001) found not a single mention of the term. There is a citation to "action and agency" in Bottomore's *Dictionary of Marxist Thought* (1983), but the text refers exclusively to action, and there is no mention of agency. Nor is there any entry for agency in the *Great Soviet Encyclopedia*.

A thorough historical treatment of agency is beyond the purposes of this chapter. Beloved by some Marxist thinkers, especially in the recent period, agency in my view is akin to free will in Christianity. Free will offers the possibility of attainment of the kingdom of heaven if individuals act for the good. For Marxists, at least those in the classical mode, the revolution is made *not* because individuals will it but because the dynamics of class struggle lead to the confrontation between proletariat and bourgeoisie. Just as people believe the exercise of free will for the good manifests itself in the kingdom of heaven, the class struggle produces the socialist revolution.

The attainment of revolution, Marx and the Marxists realized, was embodied in the distinction of the proletariat as the class *in* itself and the class *for* itself. When the proletariat failed to undertake its revolutionary mission, Marxist theory had to be adjusted by explanations such as false consciousness or poor leadership of the working class. That was in the "classical" period of the rise of the proletariat. By the 1950s, prescient social observers such as Daniel Bell (1953) recognized that the blue-collar base of the working class—that segment that classical Marxists held would continue to increase in size and power—was no longer growing, and that union membership had peaked. C. Wright Mills saw the changes (although he treated them differently) when he wrote *White Collar* (1951), arguing that the rise of the white- and pink-collar proletariat would function differently from the older blue-collar proletarians.

By the mid-1960s, although it took time to soak into the consciousness of classical Marxists, most blue-collar proletarians were supporting the Vietnam War. Such revolutionary dynamics as existed had moved on to youth, students, civil rights and antiwar activists, farm workers, and feminists. Single-issue identity politics emerged, and with it the demise of revolutionary expectations. This was reinforced by the decline and collapse of the Soviet Union and the turn to capitalism in the ostensibly socialist countries of Eastern Europe.

Before turning to the consequences of these developments, it might be useful to review how the concept of agency became concretized in the movements of the classical socialist period. While Marx had witnessed the spontaneous manifestations of working-class struggles in the early formation of the proletariat, these spontaneous manifestations of agency—the search for solutions of everyday life on the shop floor, in consumption, and for political rights and cultural expression—rapidly took on coherent organized forms. For Marx, organization was critical for transforming spontaneous and sporadic manifestations of agency into organized coherent expressions. Marx dedicated his early work to theorization about the nature of capitalism, but from midlife on he became more concerned with politics and organization. Marx's recognition of organization stood in sharp contrast to the spontaneity that characterized his radical opponents, the anarchists. In other words, agency, as a spontaneous manifestation, was inadequate for the revolutionary transformation; organization was critical.

Classically, organization took two distinct forms: social democratic and Bolshevik (communist). The social democratic form saw the organic growth of a complex of political parties, trade unions, cooperatives, and cultural institutions that, it was believed, would eventually encompass state and society. Partially in reaction to the social democratic model that envisioned a gradual transformation to socialism, especially under the conditions of czarist autocracy, Lenin projected a very different organizational format, arguing that organic growth would be insufficient for socialist transformation. A politically coherent cluster of revolutionaries was required to direct the diffuse manifestations of agency within the working class. Trotsky explained, "Without a guiding organization the energy of the masses would dissipate like steam not enclosed in a piston box. But nevertheless what makes things move is not the piston or the box, but the steam" (Trotsky 1937, xix).

Social democratic and Leninist approaches contain elements of agency, but both faced the problem of how to organize and sustain the continuously generated manifestations of agency. For the social democrats, the agency of the proletariat over the extraction of surplus value on the shop floor, or the agency driving workers as consumers into forming consumer cooperatives, or the agency of workers seeking political rights, became manifested in working-class parties; all these manifestations of agency required coherent organization, ultimately producing social transformation.

In contrast, the political party was central for Lenin. Workers might "naturally" gravitate toward unions because of struggles on the shop floor, or create other organizations such as were found in social democratic forms, but the party was crucial in broadening the "natural" inclination toward unionism into a more coherent struggle to capture the state (and therefore the economy and society).

So, for both social democrats and Leninists, the party—and its associated organizations—became the specific manifestation of proletarian agency. Agency was spontaneous, and it sometimes took more coherent form; the object was to make that coherence more conscious, deliberate, and continuous. One way to think about this is as serving the function of revolutionary accumulation,[3] in which resistant activity moves beyond the individual—or the isolated group—to the sociopolitical level, i.e., it takes on social movement character.

A key idea is embodied here: agency is a manifestation of dissatisfaction with conditions of everyday life and a search for solutions. While this may occur at the level of the individual, it has no social significance until that search expands to the social level. Even then it may not be significant unless it takes a form that influences the lives of many people—what Marxists used to call "the masses."

Agency: Recent Experience

Agency may be difficult to define, but, as used in this chapter, it carries an unmistakable element of counterhegemony and resistance. Six empirical examples may help clarify a conceptual approach. Three describe what I refer to as "primitive" or "undeveloped" agency, which moves from the purely individual to an increasing social level. The second three are examples of social or collective expressions of agency, where the actions of individuals are collectivized and begin to have social consequences. While these examples are distinct, the object is to present points along a continuum in which agency moves from the individual to the social level.

PRIMITIVE AGENCY

Primitive agency refers to manifestations of behavior in which individuals and/or groups shake off everyday forms of action and explore new forms—that is, become counterhegemonic—often without much consciousness, but with some level of dissatisfaction with the status quo. Only the first of these examples deals with an agrifood issue, yet all three are exemplary of primitive agency. The second and third, however, have greater potential than the first for rising to the level of social significance. One way to think about agency in its primitive form, for those looking to actor-network theory (ANT)[4] as a

3. I am indebted to Borrego (1981) for this formulation. Borrego took the concept of *capital accumulation*, an economic concept, and formulated the concepts of *capitalist accumulation* and *revolutionary accumulation*. These are political, ideological, and organizational forms of accumulation.

4. Actor-network theory (ANT), currently the rage among many social scientists, is based on an old and fundamental notion in the social sciences that people are organized in networks. See Latour (1987, 1988, 1993).

theoretical system, is to consider situations where actors are dissatisfied with an existing hegemonic system, seek a way out, but find no network within which to *enroll*, or refuse to join such networks where they exist.

Bill and rBST. I went through a personal experience of rejecting a hegemonic development with the introduction of recombinant bovine somatropin (rBST) when it was introduced in milk production. At the time, dairies that refused to use rBST were prohibited by a government agency from advertising that their milk was free of rBST. I was sufficiently incensed that I stopped buying cow's milk and shifted to soy milk. This was an act of personal dissatisfaction. There was an anti-rBST network in which I could have enrolled. I followed that movement until I got bored, but I never joined. But I never returned to cow's milk, either.[5] Thus, my agency in rejecting cow's milk was personal—it made me feel good—but it had no social consequences. This individual action *might* have had some social consequences, for example, an increased orientation toward organic foods (see below), but, as I experienced it, I was not "enrolling" in any resistant network as much as expressing a personal reaction.

Madame Lacroix and the Tricolour. A *New York Times* article (3 March 2003, A28) on fashion showings in Milan before the Iraq war describes fashion expressions in a period of crisis:

> During the Second World War, Mr. Lacroix [the fashion designer] went on, his mother was a girl of 16 living in occupied Arles. To signal her own resistance, she incorporated a fragment of color from the forbidden French flag in her clothes every day. "A little bit of blue, red or white in each outfit," Mr. Lacroix said, adding that if there was anything that decades in the design world had taught him, it was that symbols, however small, can sometimes surprise you with their weight.

Had increasing numbers of French citizens behaved similarly and hundreds of thousands of French men and women showed a little bit of blue, red, and white, this might have risen to the level of social significance as a form of protest and probably would have stimulated Nazi reprisals.

Women's Hair in Islamic Iran. A report from Iran in the *New York Times* (7 February 2003, A4) tells about "a guerilla struggle [that] rages on between

5. Full disclosure: I continued to eat cheese and yogurt made from cow's milk. Although I personally did not participate, reaction to the government's action rose to the social level and the policy had to be amended.

women who want to show their hair and conservative elements determined to preserve what they see as Islamic purity. Hair, in short, has become a measure of resistance to the forced will of the Islamic Republic." A beautician is quoted as saying that "forcing women to hide it [their hair] is just part of a bigger power struggle against women. My scissors and combs are my weapons in fighting for the four strands of hair women can show." The article notes that while men "grow Islamically correct beards," they are supposed to keep the hair on their heads short, and "some men have grown their hair shoulder length and longer." The struggle over hair is also an expression of class in Iran: "Both male and female lower-class morals police tend to single out middle-class or foreign women." This example involves agency moving from an individual to a collective level.

Several points should be emphasized. First, it would be a mistake to underestimate the social importance of symbols, even if they are relatively inconspicuous or hidden. Scott (1985) has shown that there are "weapons of the weak" that are manifested as "symbolic sanctions: slander, gossip, character assassination" (25), or even more expressively as "foot dragging, dissimulation, false compliance, pilfering, feigned ignorance, slander, arson, sabotage" (29). Second, unless such behavior rises to a more coherent social expression, its manifestation remains individual; there has to be some way to *accumulate* such expression in the form of a network or organization.

These examples represent the potential for antihegemonic expressions of agency, although none has had such an outcome. And this poses a problem for those of us trying to understand agency better: what are the conditions under which such individual expressions transcend the individual level and rise to the social? For, notwithstanding the significance of such individual— or group—expressions of counterhegemony, the question remains: do such expressions portend a later emergence as social movements? If not, such expressions are interesting, but not particularly significant in a social sense.

AGENCY AS A COLLECTIVE PHENOMENON

Other examples illustrate movement from individual to social levels, that is, when individual reactive behaviors take on social significance. The cases are the explosion in organic consumption as a consequence of three environmental crises in 1989, particularly the reaction to the use of the agricultural chemical Alar; the United Farm Workers (UFW) boycott of table grapes in 1971–73; and the international boycott of Nestlé over infant formula (IF) promotion in Third World settings in the 1970s and 1980s.

Organics Jump the 1 Percent Ceiling: Alar and Environmental Crises. For years, despite the impact of Rachel Carson's *Silent Spring* (1962) and the burgeoning environmental movement, organic foods languished as a minuscule percentage—less than 1 percent—of total U.S. produce consumption. Consumers were uncertain whether claims that certain foods were free of pesticides and other chemicals were accurate, or that organic food would be beneficial. There was no registration or certification. Supply was uncertain. And frequently organic promulgators were hippie "deviants" operating outside the mainstream.

Each time there was an environmental horror story, there would be a bump upward in consumption that then leveled off; organics continued to be a minuscule segment of consumption. In March and April 1989, three national environmental crises unfolded, so that organics jumped the 1 percent ceiling and began to grow approximately 20 percent annually.

This is not the place to set out the history of the acceptance and growth of organics (see Guthman 2004). As the three crises unfolded, U.S. consumers became alert to the need to protect themselves. In effect, agency, as a defensive force against a dominant production system, became broadly clear; millions of individual consumption decisions produced an important outcome in agricultural chemical usage.

The crises involved Alar, a chemical used in apple production for cosmetic purposes; the discovery of two cyanide-contaminated grape berries (not bunches) in a shipload of grapes from Chile; and the *Exxon Valdez* oil spill in Alaska. Alar threatened consumers of apples and apple products, i.e., a large percentage of the population. The contaminated grapes baffled everyone, because they raised the question of how inspectors could find two berries in a shipload of grapes. The "discovery" led to the destruction of all grapes on supermarket shelves, back to the vineyards in Chile, and all the ships and trucks in between. The *Exxon Valdez* story focused on a delinquent captain whose ship ran aground, causing the largest oil spill in history up to that point. All three were extensively covered in most of the national press.[6]

---

6. The *New York Times* coverage of the three crises provides an indication of the national coverage. On Alar, although there had been several stories before the crisis broke, the *Times* carried reports on March 16, 17, 21, 24, 26, 30, April 5, and May 7, 12, 14, 16. Reports on the grapes were published on March 14–19, 21–23, 25, 27, and April 16. *Exxon Valdez* stories appeared between March 25 and April 9. On March 4 there was a report on another Exxon tanker spill in Oahu, Hawaii. Having nothing to do with these three crises, the situation was not helped by the discovery—and destruction—of four hundred thousand contaminated chickens in Arkansas, reported on March 14; and three stories on March 21 on an EPA call for a ban on the use of aldicarb—a toxic chemical found in Great Lakes birds—used on potatoes and bananas.

While most dramatic from a national perspective, the *Exxon Valdez* spill and the other crises brought home to U.S. consumers their vulnerability to agricultural chemicals and pesticides. Agency, in this case, manifested itself in two variants of the NIMBY (not in my backyard) phenomenon: NIMB (not in my body) and NIMBB (not in my baby's body). If, as a consumer, there is little you can do to protect yourself in general, once alerted to a specific danger such as Alar, there is something you can do: avoid apples and apple products. To the chagrin of the apple industry, consumption of fresh apples dropped substantially within a week. As reported in 1991, Washington apple income dropped from $875 million to $745 million, a loss of $130 million, after the story broke (*Packer,* 26 January 1991, 3A).

The Alar story got broad coverage in the *Packer,* a trade newspaper of the fresh produce industry, which reported details not covered in the *New York Times.* The *Packer,* in the fall of 1986, reported that the EPA had equivocally banned and then lifted the ban on Alar. Although there was pressure from supermarket retailers for a ban, Uniroyal, the maker of Alar, insisted the chemical was safe, based on a study of six hundred rats. During 1987 and 1988, controversy continued over Alar, with segments of the apple industry arguing for its banning. The Natural Resources Defense Council, an environmental organization, reported on 25 February 1989 that children were at risk of cancer because of exposure to carcinogens including Alar. The same issue of the *Packer* reported that Meryl Streep, the actress and mother of three young children, had spoken to the Junior League in Montclair, New Jersey, about Alar's danger. The following day (reported in the 4 March issue of the *Packer*), *60 Minutes* did a segment on Alar for a national audience. The television report appeared simultaneously with a National Academy of Sciences report on the benefits of eating fresh produce. Also on 4 March, the *Packer* reported that apple sales were slowing and that schools had become wary of making apples available to students. Meryl Streep appeared on the Phil Donahue show on 6 March, again talking about the dangers of Alar.

On 25 March the *Packer* reported that Streep was forming an organization, Mothers and Others for Pesticide Limits. This occurred simultaneously with the breaking of the story on the cyanide-contaminated Chilean grapes. On 1 April, Consumers Union weighed in with a report that three-fourths of apple juice and 55 percent of raw apples carried Alar residue. By 8 April, the *Packer* was reporting that apple growers were growing frustrated with the situation. With consumption of fresh apples collapsing, consumers also discovered that Alar might turn up in apple juice and other apple products. Since applesauce is a major baby food item, its consumption began to drop.

The upshot was that apple growers pressured Uniroyal to stop producing Alar. Though the company resisted at first, Uniroyal caved in as far as U.S. production was concerned. With a considerable amount of apple juice concentrate being imported from other countries, the spontaneous boycott of apples and apple products finally drove growers to insist that Uniroyal cease producing all Alar, and Uniroyal finally complied.

Because these three incidents occurred in a short period, the U.S. population was alerted to the dangers of a particular agricultural chemical used on a very specific—but widely consumed—agricultural commodity. Consumers got Uniroyal to cease producing Alar only secondarily; with growers suffering from the spontaneous national boycott, they were the ones who pressured Uniroyal directly.

The cyanide-contaminated grapes and the *Exxon Valdez* crises amplified consumer resistance. There was little or nothing consumers could do in either situation, but there was something that all could do as far as Alar was concerned: stop buying apples and apple products.

The agency manifested in the Alar case was essentially *personally defensive* rather than *actively resistant*. This form of agency, however, was devastating to the apple growers, whose only recourse in reestablishing consumer confidence in apples was to get Alar completely eliminated.

There was, however, another defensive measure consumers could take: shifting from conventional to organic produce. This shift became immediately recognizable to the produce trade, as reported by the *Packer*.

Ten years earlier, in 1979, in response to pressure to pay more attention to "alternative agriculture," the U.S. Department of Agriculture (USDA) had organized a panel of experts that suggested that the potential of organics warranted attention. The *Packer* (26 July 1980, 16A) greeted the suggestion by urging caution; the *Christian Science Monitor* (13 August 1980) reported optimistically that "organic farming blossoms into favor at U.S. Agriculture Department." In general, the *Packer* was negative about the prospects of organic farming. Occasionally an article would report some event involving organics, but on the whole, the trade—particularly as manifested by the newspaper—refused to take organics seriously. Indeed, immediately before the Alar story broke, the *Packer* ran a story under the headline "Organic Demand Weakening, Wholesalers Say" (13 March 1989, 13B).

The situation changed dramatically with the Alar affair. The 25 March 1989 issue (5A) ran a headline "Organics: Hot Demand, Short Supply." To give readers a better sense of how demand had increased, a short article reported the effects in Larry's Markets in Seattle, which ran an advertisement for organic apples that happened to appear immediately following the *60 Minutes* report on Alar. The markets' director of produce operations reported, "We went from selling 600 to 700 pounds [of organic apples] a week to selling 6,500 pounds."

Summing up the events of 1989, Stephen Pavich, a large organic-table-grape grower, commented:

> The interesting point about the last few years is the way in which public awareness of organic produce has risen. And this is linked to a series of environmental disasters in Europe and the USA. . . . The nuclear power station accidents of Three Mile Island and Chernobyl, the personal and ecological tragedies of Sandoz and Bhopal. . . . In the USA . . . the issue of food safety has also made headline news. . . . The Alar and Chilean grape stories were probably the biggest, but back in 1985 we had a watermelon poisoning scandal in California and in other states which really made a lot of people sit up and think. (*Eurofruit,* December 1989, 34–35)

The consequences of the three events become clearer when an assessment was made of organic food sales from 1980 to 1990. In 1980 annual sales of organic produce were valued at $175 million. By 1985 sales had risen to $435 million, by 1988 to $893 million, by 1989 to $1.25 billion, and by 1990 to $1.5 billion (*Packer,* 23 March 1991, 1A).

There have continued to be ups and downs since then. By September 1989, the new plateau of consumption was noted by the *Packer* headline "Organics' Strength Wanes at Retail; Look, Price Blamed" (2 September 1989, 14B). Even so, through the 1990s and continuing in the 2000s, the sale of organics has grown 20 percent annually.

The 1989 crisis represents an example of what I call *latent, defensive* individual agency becoming social. By *latent agency* I refer to a growing sensitivity in the general population to food and environmental issues that becomes social by developing a focus—for example, on apples, about which individual consumers could take individual action. *Defensive agency* refers to a specific action intended for protection of self or those immediately proximate, especially children. When these individual actions accumulated sufficiently, and when growers could find no other way out, they became the actors, pressuring Uniroyal to cease production of Alar completely; individual actions had become impressively social. This example is notable in that the crisis did *not* lead consumers to form organizations.

The UFW *Table Grape Boycott.* The table grape boycott of the United Farm Workers union (UFW) used latent agency and developed consciousness, not about a defensive personal posture but to support a "good cause." Until the UFW could arouse urban consumers about the condition of farmworkers, most knew nothing about it, and therefore did not incorporate it into purchasing decisions. If the Alar boycott was purely spontaneous, the UFW grape boycott of 1968 had to be carefully, deliberately, and tediously organized. At its apogee, however, its deadly effects brought recalcitrant grape growers to the bargaining table.[7]

In 1968 a strike of Filipino grape harvesters, begun in the Coachella Valley in California, moved north into the main table grape harvesting region of the southern San Joaquin Valley. There, a still-developing union of farmworkers led by César Chávez had to decide what to do. The Mexican farmworkers joined the strike. They quickly learned that, while they could bring strikebreakers out of the fields, they could not stop grape employers from recruiting new strikebreakers. To make the strike effective it was necessary to move the strike to the point of consumption. In other words, they had to organize a national boycott of table grapes.

The boycott was successful for a number of reasons. First, in organizing the UFW, Chávez had trained members in organizational skills. This meant that he could send farmworkers, carrying authentic experiences from the field, to major urban centers to organize and "enroll" local sympathizers in the boycott. Second, Chávez was a creative organizer of media events, which gained the union continuous dramatic exposure. Chávez went on one of his fasts, and its symbolism was covered extensively by print and broadcast media. Third, with a major upsurge in political activism from the mid-1960s onward, college students became available as a renewable resource to staff the boycott organization. Fourth, grapes are fungible: other fruits can be substituted for them without causing too much pain to consumers. Fifth, the boycott could be institutionally affected by getting city councils to remove grapes from institutional consumption. Annual per capita table grape consumption dropped from five to two pounds. This forced table grape growers to the bargaining table to establish the first major labor contracts in California agriculture.

The UFW grape boycott represents the triumph of organization, capitalizing on the agency of consumers that did not exist at the outset. For the

---

7. The drama of UFW organization during the late 1960s and early 1970s generated an enormous literature. For a short summary of the grape boycott, see Majka and Majka (1982, 186–97).

boycott to become effective, it was necessary that consumers be made aware of the hegemonic aspects of California agriculture, on the one hand, and the relatively painless sacrifice, on the other, that would help an exploited group of workers. The boycott was signally successful once a counterhegemonic movement (or, more specifically, an anti–grape grower movement, or, perhaps even more accurately, a profarmworker hegemony) could be established.

In our terms, where there was no original agency, it was created by organization and proved successful. Successive boycotts in lettuce were unsuccessful because Chávez turned boycotts on and off too frequently and lettuce proved to be less substitutable than table grapes (and for other reasons).

The Nestlé Infant Formula Boycott.[8] The Nestlé infant formula (IF) boycott has elements similar to the UFW grape boycott; to be effective, the latent individual agency favoring "natural" breastfeeding had to be counterposed to the "unnatural" salesmanship of giant transnational corporations (TNCs) trading on the vulnerabilities of Third World mothers. In this sense, the agency manifested was positively associated both with "natural" norms (breastfeeding) and with resistant agency (directed against Nestlé as representative of all the IF TNCs).

There are also differences. Unlike the table grape boycott, which grew organically from a partially successful strike (it was effective in bringing workers out of the fields) but was also unsuccessful (it was unable to stop production by the continuous importation of strikebreakers), the Nestlé boycott was generated by a small number of activists in Europe and the United States.

The background of the boycott involved a growing concern in the early 1970s about the trend to shift Third World mothers from breastfeeding to IF. While Nestlé was not the only TNC to sell IF, it was the largest and the best known. Producing a huge variety of processed foods, some bearing the Nestlé label but others sold under other labels, Nestlé was at least potentially vulnerable to a boycott.

The Nestlé boycott is a useful example because it occurred in a context in which the analysis of power structure had become established in social science circles and to some degree among educated people, particularly those with an activist bent. The first gropings toward the Nestlé campaign began when a group in New Haven began research on IF and assembled a collection of eighty articles. Physicians at UCLA were conducting research on malnutrition related to IF use and had begun talking about "commerciogenic

---

8. This section is based on Hutchinson (1984) and Sikkink (1986).

malnutrition." By 1973 the *New Internationalist,* a journal in the United Kingdom, and the U.S. Consumers Union, were worried about IF. In 1974 the World Health Organization also became concerned. That same year, War on Want, an activist group in the United Kingdom, published a report, *The Baby Killer.* When it was translated into German and published by a Swiss counterpart organization, Nestlé promptly sued for defamation and libel. This had the unfortunate (for Nestlé) consequence of giving considerable publicity to the issue. In the process, the ways in which IF makers "hooked" Third World women on IF became clear: they gave new mothers free samples of sufficient volume that their breast milk would dry up; they used "milk nurses" (salespeople dressed in nurses' uniforms) to hand out samples, and so on.

In July 1976 a conference of activists from eight European countries and the United States focused on the boycott idea. A coordinating organization, INFACT, was formed in January 1977 and demanded that Nestlé and the other IF manufacturers stop distributing free samples, using "milk nurses," and promoting IF to physicians.

Nestlé and the other manufacturers refused to stop promoting IF in the Third World. The activists responded by generating boycotts in dozens of countries, calling on consumers to boycott the Nestlé label and other labels owned by Nestlé. INFACT also created a negotiating committee so that a coherent united front could negotiate with Nestlé. This involved organizations in approximately a dozen countries, no mean task. Ultimately the boycott was sufficiently painful that Nestlé and the other manufacturers agreed to negotiate. After a battle that lasted seven years, an agreement was signed in 1984 that brought the boycott to a close. (There were subsequent disagreements that need not be dealt with here since they fall outside the purview of this chapter.)

The agency manifested in the Nestlé boycott reflected the normative claim of the superiority of breast milk, on the one hand, and a rejection of the callousness of the giant TNCs, on the other. Both aspects lay essentially latent in the hundreds of thousands of consumers who ultimately joined some aspect of the boycott.

Theorizing Agency: Toward an Empirical Typology

The six examples discussed above provide a basis for moving toward a level of abstraction in seeking to develop an empirical typology of agency. I want to summarize several elements that should be kept in mind in conceptualizing agency.

First, agency always develops in a particular context. The context may be dramatic or explicitly acknowledged socially, e.g., the Nazi occupation of France or an accumulation of environmental crises. The context may also be "latent," "buried," or not explicitly acknowledged: fundamentalist Islamic injunctions about hair, the plight of farmworkers, breastfeeding of babies.

Second, there is a significant distinction between individual and collective manifestations of agency; similarly, some manifestations of agency have a lesser or greater potential for being resistant. Table 2.1 sets out these differences.

Table 2.1 indicates the distinction between the first three examples and the following three in terms of social consequences. This does not mean that the first three did not have potential social significance. If citizens in occupied France had noted Mme. Lacroix's colors and had begun to reproduce them for themselves, and this movement had spread, it potentially could have had social consequences. Similarly, Bill and rBST had no social significance. But the accumulation of grievances with the way food had been handled by the USDA later contributed to the organized-spontaneous reaction when the USDA promulgated its first draft of organic food regulations. After the

Table 2.1   Agency: Case Examples

Case	Context	Action	Consequence(s)
Bill and rBST	USDA prohibition of rBST-free labels	Boycott cow's milk, shift to soy milk	None
Mme. Lacroix	Nazi occupation–France	Red, white, blue colors worn	None
Iranian hair	Female repression	Hair coloring, etc.	Probably none
Alar	Environmental disasters	Spontaneous boycott of apples and apple products	Withdrawal of Alar; organics jump over 1%
Table grapes	Farmworker strike; 1960s activism	Organized grape boycott	Collapse of grape sales; union recognition
Infant formula	Third World marketing	Transnational boycott	IF manufacturers stop marketing in Third World

department rejected the recommendations of its own advisory board by permitting certification using procedures that are anathema to organics, organics advocates organized an outpouring of correspondence that forced the department to withdraw its draft. In this case, the USDA became the focus of consumer animus. Individuals are more likely to express their agency when a specific focus is identified than they are when there is only a vague or indistinct focus.

Yet another way to think about these six examples is to consider them in the context of two variables: whether the expression of agency is spontaneous or whether it manifests some degree of organization and has some potential for resistance. Figure 2.1 sets out the cases against these two variables.

Figure 2.1 contains implications for what I call *resistant agency*. Individuals and collectivities, whether consciously or not, may express agency in ways that pose no threat to existing ideas and practices, or in ways that reject existing hegemony. Every day consumers are confronted by new choices, new

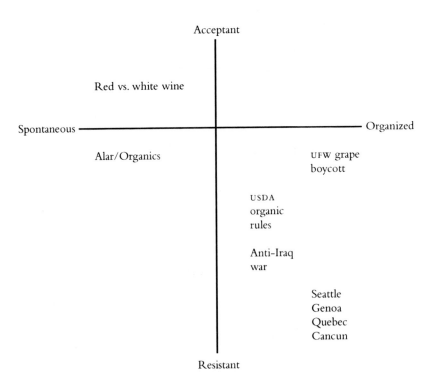

Figure 2.1   Agency: Resistance—Spontaneity/Organization

products, new ideas. A few may be accepted by some consumers, but many will perish by the wayside. Those that are accepted or become wildly popular are expressions of agency but probably pose no issues in consumers' minds. In the 1970s, for example, when the "wine revolution" was getting under way and U.S. consumers began to consume wine in significant quantities, winemakers were convinced that their choice would be red wines. When consumers confounded this expectation, there was no challenge to the wine "revolution." Was this an expression of agency? Consumers were making individual decisions that were sufficiently profound to rise to the social level, but the collective impact can be considered counterhegemonic only in that they rejected the expectations of the wine producers. Winemakers had to grub up vines or graft them to white wine varietals, but there was no fundamental change in the wine system.

Finally, it is useful to consider agency in terms of the degrees of resistance and the degree of organization. This places the various examples along a continuum. The red versus white wine issue made no challenge to any ideas and created no problems in society (although it certainly did for wine grape growers, winemakers, and wine marketers). This was a spontaneous, unorganized expression of agency, as was the Alar-organics example. While organized activity (Meryl Streep's public stand, the *60 Minutes* segment) triggered a spontaneous boycott, the boycott focused on a specific, identifiable, and fungible product (apples), but also was an expression of dissatisfaction with the use of chemicals in food production, still only a moderate manifestation of resistance. The four examples shown in the "resistant-organized" quadrant in Figure 2.1 manifested differential degrees of organization and resistance. The UFW table grape boycott was highly organized but only moderately resistant. Major events such as the Iraq war have generated much higher levels of resistance. Highly organized and resistant activities are exemplified in popular reactions to meetings of the Group of Seven or the World Trade Organization, where the leaders of the "free world" either meet behind physical barriers (as in Seattle, Genoa, Quebec, Cancun) or withdraw to isolated locations impossible for protestors to reach (Doha, Qatar, or somewhere in remote western Canada). This also suggests pursuing the spontaneous-organized dimension of agency somewhat further, raising the issue of whether agency produces *continuing* organization. The single case of the infant formula controversy is cited, where INFACT shifted the focus of its attention to General Electric and the general issue of corporate abuse. The details of the shift cannot be detailed here, but the organization continues its activities at the time of this writing.

From Resistant Agency to Social Movement: Assessing the Potential

The six cases—three *small-scale, individual,* or *partially social* and three that rise to the *social* or *social movement* level—were selected because they illustrate the marginal area between *precursor* or *protosocial movements* and *genuine* (if limited) social movements. While none of these six cases can or should be ignored, it is obvious that the first three did not rise to the point of social significance. Wearing blue, white, and red discreetly during Nazi occupation, showing a maximum of hair in a repressive Islamic regime, and individually drinking soy milk rather than cow's milk are hardly revolutionary acts. They can be satisfying to the individuals involved: at a social-psychological level they can be manifestations that individuals have not given up, but are continuing a struggle surreptitiously, employing "weapons of the weak" (Scott 1985). They become the potential feeding ground for counterhegemonic action or, to use ANT terminology, they provide opportunities for individuals to enroll in actor networks.

Those of us interested in counterhegemonic movements ignore such individual and small group manifestations of counterhegemony at our peril. Our problem remains that we need better ways of assessing such protodevelopments and how to work with them to become better organized. To use two very different languages, from a Marxist point of view, we need to develop instrumentalities of revolutionary accumulation, or, to use actor-network terminology, we need to create better linkages in the networks to include more individual and group actors.

The two theoretical systems have treated this problem very differently. For Marxists, agency was historically manifested and accumulated through the party; for devotees of actor-network theory, agency is manifested through spontaneous action or, as Tanaka, Juska, and Busch (1999) have shown in the case of rapeseed, through the organized intervention of the state, in which case the actor network is hardly counterhegemonic. ANT, in effect, is nonrevolutionary, whereas Marxist agency maintains social change or revolutionary expectations, although with neither a party nor the proletariat as a *deus ex machina.*

It is because of those two lacunae that Marxist intellectuals have fallen back on agency, finding in its dialectic the generation of counterhegemonic tendencies. What is missing, however, is an instrument of accumulation equivalent to the party.

In 1902, Lenin posed the problem *What Is to Be Done?* in the context of a growing proletariat, an overwhelming but stupefied peasantry, and a handful of revolutionary intellectuals who believed czarism could be overthrown. Lenin proposed as his solution the cadre organization that later came to be known as Bolshevism. Both social democratic and Bolshevik models represented instruments for revolutionary accumulation.

Today, that kind of revolutionary transformation is virtually inconceivable; practically no one talks "revolution" anymore in Western democratic capitalist societies.[9] At the same time, movements continue to bubble up from the grass roots to the point that global leaders have to meet in isolated locations or hidden behind police-fortified barricades.[10]

Readers will see the variations in agency and the relationship of agency to structure in the chapters that follow. The chapters by Johnston and Shreck are explicit in recognizing the importance of counterhegemony in defining agency. Shreck also emphasizes the notion of *levels* of agency in her description of the Fair Trade movement, which is counterhegemonic but far from transformative in its goals. Skladany's chapter, on salmon, is less about agency per se and more about economic interests (farmed salmon versus fished salmon), with a moral or normative interest augmenting the agency of salmon fishermen. The chapter by Munro and Schurman on the anti–genetic modification movement illustrates how a technically based network has successfully challenged the hegemonic pro–GM science network in what was originally a technically based agency of individuals that became collectivized and rose to the social level, but has not yet created a social movement. Guptill's examination of Puerto Rican organics illustrates individual expressions of counterhegemonic agency that have failed to rise to the social level. Jussaume and Kondoh examine the agency of producers and consumers in four Washington State counties with respect to direct purchasing of agricultural products. And finally, focusing on agency as acts of resistance, Tanaka and Ransom show how different social strata in New Zealand and South Africa operate in the red meat system. The range of empirical cases offers testing grounds for the meaning of agency and its utility as an analytic concept in examining social change.

In recent years we have experienced a period in which identity and single-issue politics have demonstrated the capacity to move hundreds of thousands of people to express their agency in action, but where the fissiparous

---

9. "Revolution" in the political sense of Marxism. In contrast, there are daily "revolutions" in shaving cream, theater, washing machines, etc.

10. There is considerable irony here. In the classical period of revolutionary Europe, it was the revolutionists who threw up the barricades; now it's the state, to protect national leaders from, on the whole, peaceful protesters.

narrowness of particular ideologies has precluded the generation of any accumulation instruments. Agency exists and will continue to exist, but until such instrumentalities can be invented, agency will continue to manifest itself in relatively circumscribed and limited and discontinuous social movements. Revolution is no longer on the agenda, but social change remains as important as ever. We need an updated Lenin, appropriate to our times and the socioeconomic circumstances of our lives, to write our *What Is to Be Done?*—that is, to propose accumulation instrumentalities that can capture the agency centered on progressive social change that is engendered continuously, and to make that agency a more effective instrument. Whether such a development is possible remains to be seen.

REFERENCES

Bell, Daniel. 1953. "The Next American Labor Movement." *Fortune,* April, 120–23, 201–6.

Borrego, John. 1981. "Metanational Capitalist Accumulation and the Emerging Paradigm of Revolutionist Accumulation." *Review* 4 (spring): 713–77.

Bottomore, Tom, ed. 1983. *A Dictionary of Marxist Thought.* Cambridge: Harvard University Press.

FitzSimmons, Margaret, and David Goodman. 1998. "Incorporating Nature: Environmental Narratives and the Reproduction of Food." In *Remaking Nature: Nature at the Millennium,* ed. Bruce Braun and Noel Castree, 194–220. London: Routledge.

Giddens, Anthony. 1984. *The Constitution of Society: Outline of the Theory of Structuration.* Berkeley and Los Angeles: University of California Press.

Guthman, Julie. 2004. *Agrarian Dreams: The Paradox of Organic Farming in California.* Berkeley and Los Angeles: University of California Press.

Hutchinson, Susan. 1984. "An Examination of the Artificial Infant Formula Controversy and the International Organizing Campaign Waged Against the Infant Formula Industry." Senior thesis. University of California, Santa Cruz, Department of Community Studies.

Jasper, James M. 1997. *The Art of Moral Protest: Culture, Biography, and Creativity in Social Movements.* Chicago: University of Chicago Press.

Latour, Bruno. 1987. *Science in Action: How to Follow Scientists and Engineers Through Society.* Cambridge: Harvard University Press.

———. 1988. *The Pasteurization of France.* Cambridge: Harvard University Press.

———. 1993. *We Have Never Been Modern.* Cambridge: Harvard University Press.

Lenin, Vladimir Ilyich. 1902/1943. "What Is to Be Done?" In *Selected Works,* 10 vols., 2: 25–192. New York: International Publishers.

Majka, Linda C., and Theo J. Majka. 1982. *Farmworkers, Agribusiness, and the State.* Philadelphia: Temple University Press.

Marx, Karl, and Friedrich Engels. 1975–2001. *Collected Works.* 50 vols. Moscow and New York: Progress Publishers and International Publishers.

Mills, C. Wright. 1951. *White Collar: The American Middle Classes.* New York: Oxford University Press.

Parsons, Talcott, and Edward A. Shils. 1951. *Toward a General Theory of Action.* Cambridge: Harvard University Press.

Scott, James C. 1985. *Weapons of the Weak: Everyday Forms of Peasant Resistance.* New Haven: Yale University Press.

Sikkink, Kathryn. 1986. "Codes of Conduct for Transnational Corporations: The Case of the WHO/UNICEF Code." *International Organization* 40 (4): 815–40.

Tanaka, Keiko, Arunas Juska, and Lawrence Busch. 1999. "Globalization of Agricultural Production and Research: The Case of the Rapeseed Subsector." *Sociologia Ruralis* 39 (1): 54–77.

Trotsky, Leon. 1937. *The History of the Russian Revolution.* New York: Simon and Schuster.

INSTRUCTOR'S RESOURCES

Key Concepts and Terms:
1. Agency
2. Primitive Agency/Collective Agency
3. Social Movements/Protosocial movements

Discussion Questions:
1. Compare and contrast the three different types of agency detailed in this chapter: individual, latent, and defensive.
2. What is the role of context in understanding the exercise (or the lack thereof) of agency?
3. Iranian women style and expose their hair in ways that express their individuality but also their resistance to social control. Can you think of other ways in which personal symbols such as hair have been used to counter hegemonic structures?

Agriculture, Food, and Environment Video:
1. *Fighting for Our Lives: The United Farm Workers 1973 Grape Strike.* National Farm Workers Service Center, 1974 (60 minutes).

Agriculture, Food, and Environment on the Internet:
1. United Farm Workers: www.ufw.org/.
2. The Nestle Boycott: www.breastfeeding.com/advocacy/advocacy_boycott.html.
3. Corporate Accountability International (formerly INFACT): www.stopcorporate abuse.org/.

Additional Readings:
1. Guthman, Julie. 2004. *Agrarian Dreams: The Paradox of Organic Farming in California.* Berkeley and Los Angeles: University of California Press.
2. Mooney, Patrick H., and Theo J. Majka. 1995. *Farmers' and Farm Workers' Movements: Social Protest in American Agriculture.* New York: Twayne. (See especially chapter 6, "The United Farm Workers Era," on the UFW boycotts.)
3. Sikkink, Kathryn. 1986. "Codes of Conduct for Transnational Corporations: The Case of the WHO/UNICEF Code." *International Organization* 40 (4): 815–40. (Sikkink examines the INFACT boycott.)

3

RESISTANCE, AGENCY, AND COUNTERWORK: A THEORETICAL POSITIONING

Norman Long

The main task of this chapter is to unravel some of the conceptual and interpretative problems implicated in studying issues of "resistance" and "agency." But why should we take up these seemingly old intellectual chestnuts when there already exists a voluminous body of literature on these matters?

My response would be that, after the invention of the "cultural dope" of structuralism, the "calculating utilitarian" of rational choice theory, and the "death of the subject" of postmodernism, we need to reaffirm once again the importance of issues of agency linked to the organizing practices, reflexive capacities, and creativity of social actors—even those who have to answer to hegemonic authority. In her book *Being Human: The Problem of Agency,* Margaret Archer (2000, 2) comes to a similar conclusion. She writes, "So now it is our job to reclaim Humanity . . . [from] the Academy, where strident voices would dissolve the human being into discursive structures and humankind into a disembodied textualism." She is right. We must return to an understanding of the fundamental sources and trajectories of change and heterogeneity, but this time without the shackles of class categories or the hegemonic script of globalization.

This chapter, then, pursues this line of argument through a critical assessment of several key concepts relating to issues of resistance and agency. What I have to say, I believe, is generally important for how we go about understanding the dynamics and outcomes of social struggles within today's global society. One striking field of contest, of course, involves the momentous transformations now taking place with regard to agricultural and food practices. Here we encounter a whole range of battlefields focusing on inequities in international trade agreements, unacceptable levels of labor exploitation, controversies concerning the role of science, GM crops, and methods of controlling environmental pollution, as well as the implementation of bureaucratized systems for the measurement and regulation of product quality and food safety. These struggles involve an enormous multiplicity of social actors and institutions connected in diverse ways.

While the following chapters illuminate, through specific cases, the dynamics of such "fights over food," here I argue the advantages of adopting actor-oriented concepts and theories for probing questions of resistance, agency and structure. My general theoretical position differs, therefore, from that of Bill Friedland in the previous chapter, whose perspective is more structuralist and neo-Marxist in character. At the same time, however, I endorse the importance he places on distinguishing between different modes of agency and their intended and unintended outcomes.

How to Approach "Resistance"

The notion of resistance is commonly used to characterize struggles against sites of power and authority, or what are often called "dominant" actors or regimes, carried out by less powerful or so-called subordinate actors, both individual and collective. Resistance is therefore said to be a feature of the type of hierarchical conflict that erupts between peasants and landlords, workers and managers of enterprises (private or state-owned), small-scale farmers and moneylenders or intermediaries, land squatters and private or public property owners, and, more generally, citizen groups contesting, directly or indirectly, the authority and/or policies of the state or powerful international corporations. While the nature and effectiveness of these different scenarios of resistance will vary considerably, the question remains whether the word "resistance" is apt or robust enough to embrace fully the complexities involved in these struggles.[1]

This problem is addressed by Charles Hale (1994) in his history of the struggles between the Miskitu Indians and the Nicaraguan state. He suggests that in order to capture the complexities and subtleties of this process, one

---

1. The distinction is sometimes made between "active" and "passive" resistance. It would be wrong to draw too sharp a difference between these, however, since "passive" resistance of the kind illustrated by sit-down strikes or silent demonstrations is every bit as active or engaged as any other move against authoritative bodies or power holders. This is also true of those everyday acts of defiance that individual subordinates may carry out, such as failing to take notice of orders or rules or feigning ignorance of them, foot dragging, or delaying doing certain tasks, playing language games that ironically question status hierarchies, or the gentle sabotage of plans or procedures that James Scott (1985) so aptly describes as "weapons of the weak."

has to avoid equating resistance with simple or untrammeled opposition to the dominant order. As he perceptively puts it, though I have added a few phrases to tone down the overly culturalist flavor of his argument:

> Responses to oppression [or domination] are multivalent, combining rejection with partial acceptance, resisting through efforts to appropriate and subvert the cultural symbols [as well as the strategies and policies] of the dominant order. As a result, they create a cultural [and organizational] form, in part actively resistant, in part expressing the "common sense" premises that come directly from dominant actors and institutions, and in part consisting of symbols [and practices] whose meanings remain undefined, ambiguous. (25)

The Social Embeddedness of Resistance and the Issue of Hegemony

This provides us with a first glimpse of a way forward in the debate on resistance and agency. The message is clear: we need to develop a theoretical perspective that enables us to explore the social embeddedness of specific acts of resistance, just as one would do for other forms of social action, including the assumed opposite of resistance—"quiescence."

By "social embeddedness" I imply more than the kinds of self-organizing processes, interpersonal networks, and informal normative commitments identified by Granovetter (1985) as essential for accessing resources, developing livelihood strategies, and managing enterprises or projects. Although such dimensions are obviously critical for the kinds of mobilizing and organizing processes necessary for implementing acts of resistance, their precise form and content, as well as the sociopolitical spaces they carve out or seek to occupy, are shaped by a broader set of conditions and components. Thus we also need to specify the ways in which existing institutional constraints, knowledge/power processes, and material affordances[2] shape the possibilities

---

2. Gibson (1979) first introduced the term "affordance" to argue against the assumption of Gestalt psychology that the perception of an object is determined by the needs or desires of the observer. Instead, Gibson emphasizes that perceived "objects" and people's environments in fact possess both "objective" and "subjective" properties. More recently, Hutchby has used the notion of affordance to explore how we interact with technological media—and therefore by implication with other "externalities"—that posit a range of possible "empirical" uses, while at the same time we shape both the cultural and interactional properties of these technologies. This type of analysis aims to avoid the extremes of social constructivism, which asserts that the meanings of particular technologies, material objects, and social arrangements are simply anchored to the range of discourses accorded them

for engaging in particular actions, counteractions, and discourses. These elements compose the fields and arenas in which struggles take place, and are themselves reconfigured by the particular actions and negotiations that ensue. The outcomes and effectiveness of specific forms of resistance or contestation rest, then, not only on the organizing capacities and strategic capabilities of so-called subordinate actors but also on how rigid or malleable "dominant" institutional frameworks and discourses are perceived to be.

In other words, as Hale's discussion suggests, we need to probe more closely into the dialectics of "dominant" and "subordinate" social forms. This seems to imply the need for a theory of hegemony that focuses on "the entire complex of practical and theoretical activities with which the ruling class [or dominant actors] not only justifies its dominance, but manages to win the active consent of those over whom it rules" (Gramsci 1971, 244).

Such a theory addresses two critical elements: the role of institutions that regulate, control, and coerce, and the processes by which consent is created and maintained or challenged. Or, as Crehan expresses it, "the concept of hegemony serves to highlight the problematic relation between force and consent, and to draw attention to the way in which they are entwined, and how force may be concealed behind a facade of consent." Also, "how is it that certain shared accounts of the world come to be so firmly embedded within individuals' consciousnesses as to seem to those individuals part of the very texture of their own subjective being?" (1997, 29, 30).

Understanding Hegemony

A revealing illustration of how such shared accounts of the world (sometimes called social representations) acquire hegemonic status is given by Gibson-Graham (1996) in her[3] exposé of how she and her left-wing political economist colleagues unwittingly contributed to the success of the hegemonic

---

by the various actors or users involved. It does this by taking into account the potentials of specific technologies and the "preferred" readings built into them, as well as acknowledging that users can always attempt "to find a way round this attempt at interpretive closure" (Hutchby 2001, 22). Analogous debates surround the issues of how far political-economic, technological, and environmental circumstances and policies constrain the agency of farmers and other countryside actors, and to what extent the latter can create space for maneuver and novelty. See van der Ploeg (2003) for an analysis of the state of play in Dutch agriculture vis-à-vis the rise of new "expert systems."

3. There are in fact two authors, but they present themselves as a single writer. I am grateful to Magdalena Villarreal (2001) for drawing my attention to this interesting critique of political economy and for offering her own penetrating comments.

project of "global capitalism." According to her, they did this by depicting "the global economy as [composed of] sites of capitalist dominance," rather than considering the alternative possibility of depicting "economic discourse as hegemonized while rendering the social world as economically differentiated and complex" (x–xi). The latter position would, of course, have required detailed research into the interplay of commoditized and noncommoditized relationships and values, and other nonmarket factors making for economic difference and social response. Thus Gibson-Graham and her intellectual partners subscribed to the image of capitalism as "large, powerful, active, expansive, penetrating, systematic, self-reproducing, dynamic, victorious, and capable of conferring identity and meaning," in short, as a fully coherent and aggressive system that simply overpowers and devours or subsumes everything in its path. That is, it generates a "representation of the social world and endow[s] it with performative force . . . undoubtedly influenc[ing] people's ideas about the possibilities of difference and change, including the potential for successful political interventions" (x).

The same type of powerful discourse is associated with current neoliberal thinking. As Gavin Smith (2003, 19–20) explains, neoliberal discourse and practice differ from earlier economic liberalism. Whereas the latter combined a decision-making market rationality with state-provided welfare for those without resources in order "to offset class inequalities, and to fill gaps where commodity relations worked inappropriately, such as in [the fields of] health and education," neoliberalism promotes a rationality "in which social subjects are essentially no different from enterprises. Workers do possess a kind of capital—human capital—and they or their predecessors have invested in capital, producing strength and skills of course, but also love, affection, morality and so on." From here it is but a short step to rationalizing the central need for people to generate their own "social capital" based on the mobilization of certain interpersonal networks and modes of collective identity. This extension of neoliberal ideas to embrace concepts of social capital, sustainable livelihoods, and community development has in fact been vigorously applied by the World Bank and many international development aid organizations. And in this way neoliberalism has acquired a degree of public respectability and now plays a major role in generating consent, negotiated settlement, and the reaffirmation of taken-for-granted values and habits among various sectors of society. It has, that is, contributed to the formation and reproduction of new repertoires of public political and everyday culture that compose the "social person" and thus shape the ways in which people experience the world they live in.

The examples given above present the case for the role played by hegemonic processes in the production and reproduction of social and economic orders. These processes—both discursive and practical—necessarily affect how people of different social locations perceive their options and respond to them. Irrespective of whether people weigh their options carefully or simply follow well-worn paths of habit, the issues remain the same, namely, their behavior is confined in certain ways by being socialized into a given framework of rules and practices that are underpinned by existing power relations and moral authority. Hence even responses such as protest meetings, or demonstrations such as the antiglobalization campaigns, are to a degree predicated on the existence of hegemony: that is, on assumptions made about the unassailable positions and expressions of authority that demonstrators wish to challenge and bring down. Moreover, much of the "soul searching," "thinking through," and "plotting" of courses of action take place experientially during the lead-up to public protest events and are influenced both by the actions and statements made by the representatives of the "offending" multilateral organizations and governments and by how the protesters themselves assess the room for maneuver and spaces for change.

Underpinning the possibilities for certain types of resistance or strategic action, then, are particular institutional, cultural, and material components that compose the field of action. Such components, it is assumed, provide the rationale for specific types of hegemonic ideology and practice, including also those associated with the rise of "populist" forms of control and authority. That is, individuals and groups may become enrolled in certain widely propagated narratives of protest (or indeed, of preserving the status quo) that can crucially affect how they see their future choices and possibilities for action.

Yet, having reviewed these arguments, it is important to stress that we should avoid the tendency to present too structural and homogeneous a view of these processes. A far bigger challenge is that of grappling with the heterogeneity of values, interpretations, interests, relations, and models of society and morality that inevitably surround issues of power. Here a central question is how such seemingly divergent or incompatible elements are in fact accommodated, negotiated, or superseded in the processes of everyday life.

This leads me to conclude that many analyses of hegemonic processes fall short because they skate too superficially over the multiplicity and complexity of elements involved. In fact, one might even say that Gibson-Graham's self-critique of the hegemonic discourse of global capitalism stands in danger of being dismissed for being little more than a crude caricature of the phenomena in question.

Exploring the Significance of Discourse

To probe these issues and processes more profoundly requires a more sophisticated approach to matters of discourse. Discourse analysis offers a useful way of exploring the significance of particular ideas and cultural repertoires and how they *interact* and *interpenetrate* situationally (see Long 2001, 50–53). By discourse is meant a set of meanings embodied in metaphors, representations, images, narratives, and statements that advance a particular version of "the truth" about objects, persons, events, and the relations between them. Discourses produce texts—written and spoken—and even nonverbal "texts" like the meanings embodied in infrastructure such as asphalt roads, dams, and irrigation schemes, and in farming styles and technologies.

Discourses frame our understandings of life experiences by providing representations of "reality" (often taken for granted) and constitute what we consider to be the significant or essential objects, persons, and events of our world. But of course it is possible to have different or conflicting versions of the same discourse, or incompatible discourses relating to the same phenomena. For instance, discourses on development vary considerably, depending upon the political or ideological positions of the organizations or actors deploying them.

It is important, therefore, to unravel the discourses utilized in specific arenas of struggle. Discourses are not separate from social practice (Foucault 1972, 1981) and may coexist and intersect with each other. However, they are not always elaborated in the form of abstract arguments, such as we find in formal plans or policy documents drawn up by development agencies or issued by other stakeholder groups. More often, bits and pieces of discursive text are brought together in innovative ways or in strange combinations in particular situations in order to negotiate or contest certain shifting points of view. Indeed, the multiplicity and fragmentation of discourses, especially in conversational and dialogical exchanges, is more common than the clash of well-defined opposing viewpoints, beliefs, or rationalities. This also holds for the rhetorical content of official government statements drawn up by politicians and their "spin doctors."

On this basis one might argue that the advancement of any particular discourse depends on the strategic use of other discourses. This is illustrated, for example, by the way in which neoliberal policy statements, with their stress on "letting the market do its job," are now often softened by discourses that emphasize issues of "equity," "participation," "sustainability," and the problems of "marginalization." Indeed, the negative effects of structural adjustment measures eventually heralded the implementation of "social compensation" or "poverty-reduction" policies aimed at protecting the badly affected poorer and weaker social sectors. The World Bank and various national governments were obliged to introduce these to counterbalance the growing evidence of marked decline in incomes and living conditions among the poor.

Thus, in order to comprehend more fully the nature and dynamics of neoliberalism, we must allow for concepts and discourses that do not blend consistently with the prevailing "credo" of global capitalism. Such diverging points of view emerge not only within the antiglobalization lobby but also within those international organizations charged with the promotion of neoliberal policies.

> This implies taking into account the capacity of actors to respond creatively to diverse scripts at the same time, reformulating and rewriting them, translating meanings from one script to another, challenging definitions and negotiating positions, all of which involves agency, and of course responding to or subverting hegemonic practices. In the process, they might be contributing to a "consecration" of a particular script or they could subvert it. More often it is a combination of both. Thus, scripts cross-cut each other and are constantly re-transcribed, containing what one might identify as incongruities and discrepancies. (Villarreal 2001, 5)[4]

It is these disjunctures that provide the space for maneuver, even within what looks like a rigid main text.

> Labels and meanings are interpreted and reformulated in the conflation [and negotiation] of interests and the interlocking of projects[5] where a range of actors enters into play. . . . These actors are not free and self-determining individuals, nor are they entirely constituted by a pre-given script that encompasses their whole being. They display diverse forms of agency in both inducing and contesting constraints. Places in society are established and negotiated, and values and interests are given meaning and sanctioned.

---

4. Building upon arguments advanced in several earlier publications (e.g., Long and van der Ploeg 1989; Long and Long 1992; and Long 2001), Villarreal offers important insights into how the production and interweaving of scripts relate to issues of power and authority.

5. Later in the chapter I spell out what is meant by this. See also Long and van der Ploeg (1994).

This is important, since, cloaked in what has come to be identified as dominant discourse, is a simplification of whole sets of processes and negotiations, which, when observed closely, reveal the vulnerabilities of power. (5)

Later on in her discussion, Villarreal goes on to stress that

discourses are constituted through social interaction, through processes involving the interlocking and segregation of voices, and in this way are shaped by the actors themselves. The interests and projects of individuals and social groups are ensnared within the scripts, thus constituting "voices" through which to express, defend and challenge practices and discourses.

Scripts are compelling tools for power. But they are not in themselves powerful. Behind representations are agents. . . . To subvert the script, we must first identify its non-alienating voices [and] sweep them out from under the rug. We must see [for example] how women's worlds are not limited to typically capitalist circuits, and we must demythologize our own concepts of capital. Subversions to capitalism might be found within what we identify as the capitalist system itself. Studying capitalists' actions from an actor-oriented perspective, one is able to identify how even the most hard core capitalist activities are not circumscribed to what one might describe as capitalist circuits. (11)

Thus, while hegemonic ideas and relations undeniably enter into the struggles that take place over issues of authority, morality, and social control, pinpointing their precise social significance is a complex task.[6] Indeed, commencing explicitly with an interest in uncovering such hegemonic elements is already introducing a degree of reification that may act as an obstacle to detecting the frailties of authority. It is also likely to promote a "top-down" perspective that runs the risk of failing to throw much light on discourses, interests, and practices that fall outside the gaze and surveillance of so-called authoritative actors and institutions. Hence we need to develop a form of analysis that centers on understanding actors' everyday life struggles, the semiautonomous fields of action in which they operate, and the creativities

---

6. It is also unhelpful to dichotomize "dominant" and "subordinate" discourses and practices since, as Derek Sayer (1994, 377) has forcefully argued, "to abstract out, reify, and monolithically counterpose 'hegemony' and 'resistance' is to misunderstand both."

(both discursive and pragmatic) they display in resolving the problems they face.[7] Unless we attempt this, we are unlikely to identify the critical sites of struggle and the combinations of coercion and consent that compose the topography of power relations. Nor can we attain an adequate understanding of the origins and character of countermovements.

Adopting an actor-oriented perspective without aiming to capture some of the complexities and dialectics of these processes would be insufficient. It is simply too populist and romantic to assert that peasants, small-scale farmers, or consumer groups develop their own effective mechanisms for resisting the pressures of the market, transnational companies, international trade agreements, or the policies of the state. They may build their own relatively autonomous organizing practices for contesting certain regimes of control, but this is not to say that these regimes, or the ideologies and power relations that support them, do not also figure in the shaping of the types of strategic counterdiscourses and counteractions that emerge. As many studies of agrarian development have shown, even in the most confined power domains and regulated interfaces, opposing parties are often able to establish some common ground in which they can accommodate (rather than contest) each other's interests. They may also negotiate rules of engagement that enable them to conduct their affairs in a nonbelligerent manner. Interestingly, the success of such strategies often rests on the ambiguous nature of the concepts or discourses they draw upon, since this allows both parties strategic room for maneuver.[8] Furthermore, many of the strategies adopted by so-called subordinate actors in fact build upon or mimic those that have already proved effective for other, better-placed, individuals or groups.[9]

---

7. The most thoroughgoing theoretical treatment of the central importance of "creativity" in social life is that by Hans Joas (1996). He critiques "rational" and "normatively oriented" concepts of action in favor of an in-depth analysis of the "intentionality," "corporeality," and "sociality" of human action.

8. I cannot here pursue these dimensions any further. See Long (1989 and 2001, chapters 4 and 9) on social "interface" analysis.

9. I do not have the space here to review the large body of literature that offers a class-based analysis of peasant and worker resistance, but here are a few brief comments. Kearney (1996) focuses on the centrality of class struggle in his discussion of peasant resistance to landlords and transnational companies. He applauds James Scott for the attention he gives to class relations in his examination of everyday forms of resistance but questions his emphasis on the many sites where individual peasants struggle against their superiors. Kearney argues that the problem with resistance theory is that "political action is dissipated at a thousand microsites of opposition and so becomes woven into the fabric of domination without changing it. Resistance is but an aspect of perennial oppression" (1996, 155). Scott makes reference to Paul Willis's (1977) intriguing study of "how working-class kids get working-class jobs," which centers upon the paradox that in rejecting the school and its culture, the lads in fact confine themselves to blue-collar jobs. Yet Scott does not push this argument through

At this point we need to lay out more precisely the central significance of "agency."

Agency as a Root Metaphor

An abiding concern in social analysis is how to conceptualize the relationship between social action, its origins, and the forces that drive and channel it. In other words, we are confronted with issues of agency in both its individual and collective modes.

Borrowing the idea from Victor Turner (1974), we can describe agency as a "root metaphor," since it provides a condensation of certain fundamental processes in social life concerning the engagement of people with their ongoing life experiences. We can summarize the nature of agency as follows:

> Agency refers to the knowledgeability, capability and social embeddedness associated with acts of doing (and reflecting) that impact upon or shape other persons' actions and interpretations, and recursively, of course, one's own. Although agency is primarily about how actors of various types engage in actions that affect the circumstances and choices of others, agency is mostly recognized *ex post facto* through its acknowledged or presumed effects. It is persons or networks of persons that are the principal source of agency, although since objects and ideas are attributed with meanings and uses, they may also have a significant influence on people's perceptions of what is desirable and possible. Agency is composed, then, of a complex mix of social, cultural and material elements. Strategic agency signifies the enrollment of many actors in the "project" of some other person or persons. (Long 2001, 240–41)

---

to explore in detail how "resistance is integral to the reproduction of class difference" (Kearney 1996, 156). And, as Hale's study shows, the same process applies to questions of ethnic strife and identity. Of course, in adopting such an analytical standpoint, we come perilously close to assuming—rather than demonstrating—the efficacy of collective action and consciousness/ideology, whether based on class, ethnic, or other social differences. Much of this kind of literature in fact works with crude structural polarities cast in terms of the relations between so-called dominant and subordinate actors and institutions. This, I suggest, fails to grasp the important point that, however we designate the protagonists involved, each category or group will necessarily be made up of a highly differentiated set of social actors who interact with each—both across and within groupings—in diverse and incomplete ways. That is, we must recognize the heterogeneity, complexity, and changeability of these networks of relations.

We can add some clarifications to this definition. In the first place, it is important to emphasize that agency is not simply the result of some kind of extraordinary gift, mystical capacity, entrepreneurial flair, or innovative spirit possessed by talented individuals. Indeed, as my colleague and co-author Jan Douwe van der Ploeg has strongly argued, we must reject Giddens's (1984, 9) formulation that "agency concerns events of which an individual is the perpetrator," whatever his or her special qualities. "An individual only displays agency in interaction with other people or with things. . . . Even if it involves only one individual, the action expressing his or her agency should absolutely not be considered an individualistic action" (van der Ploeg 2003, 15).[10] Agency is a socially generated and a culturally defined phenomenon that takes different forms depending on the context. In some situations one might even conclude that it hardly exists at all.[11] It is for this reason that van der Ploeg (2003, 16) suggests that we need also to talk about "non-agency" in the sense that "alongside the capability to make a difference [i.e., to shape the actions of others], the opposite, incapability, also frequently occurs. Without the latter we cannot define the former."

Central to this interpretation of agency, then, is the notion that it is embodied in social relations and can only be effective through them. The ability to influence the actions of others depends fundamentally upon there being an already existing or newly created network of ties by which a relevant actor or actors can be mobilized and enrolled to collaborate in some attempt to achieve common or at least compatible ends. Social actors come in a variety of forms: individual persons, informal groups or interpersonal networks, organizations, collective groupings, and what are sometimes called "macro" actors (e.g., national governments, churches, or international organizations). However, care must be taken to avoid reification; that is, one should not assume that organizations or collectivities, or categories based on class or other criteria, act in unison or with one voice. In fact, collective and organizational endeavors are better depicted in terms of coalitions of actors, interlocking actor projects, and networks wherein there is always some dissent and negotiation concerning particular strategies or rationalizations.

A further important dimension of agency is its connection with perceived future scenarios and evolving social relations. The future is, as it were, filled

10. This is a critical issue that I discuss in Long and Long (1992, 25–26): "It is at this point that the individual is, as it were, transmuted metaphorically into the *social actor*, which signifies the fact that social actor is a social construction rather than simply a synonym for the individual."

11. This is why it has been proposed that we need several other words to cover these other eventualities; i.e., "agents" can be contrasted with "patients," "stakeholders" with "bystanders," and so forth.

in in diverse ways by the different images and strategic expectations of the actors concerned. Hence it is not a state of affairs determined simply by existing normative frames, whether located in specific market, political, or cultural domains. Rather, it is composed of a range of divergent and often competing possibilities envisaged by the various actors. In this way agency implies ways of ordering the future as well as the present.

Such conceptualizations or representations, together with the capability to mobilize the necessary resources (material and social) for attempting to realize particular goals or states of affairs, are what we mean by "actor projects." Such "projects" are grounded in and/or reflect or translate specific sets of interests, commitments, and prospects. A decisive element concerns the organization and information-processing capacities associated with particular projects for realizing or creatively transforming visions of the future and for enrolling others in their specific projects. Understanding the contents of actor projects and how they interlock requires an analysis of the processes of negotiation, confrontation, contestation, and accommodation entailed in the various processes of mediation that take place between prospective network members.[12]

Interlocking Actor Projects and the Reconstitution of "Structure"

This leads us to consider what is often seen as the other side of the coin of agency, namely, what is usually called "structure." In Giddens's sense, structure is depicted as an organized set of elements that function to facilitate or constrain social action. That is, it is often thought of as constituting some kind of institutional or cultural framework that "rules in" and "rules out" certain behaviors, thus limiting the repertoire of possible actions. Add the element of authoritative power, and this formulation comes close to how hegemony is mostly used in the literature.

Van der Ploeg and I have argued strongly against such a conception of structure because it implies the notion of a fixed order of things that confines

---

12. Compare this account of interlocking projects with that of Bruno Latour's (1994) on technical mediation. Latour, whose primary focus is the symmetry between human and nonhuman actors, distinguishes between four meanings of mediation: *translation*, the creation of a link between goals or agents that did not exist before; *composition*, the association of different entities in the production of a specific action; *blackboxing*, the process by which the joint production of actors and artifacts, and the contribution of spatially and temporarily distant actors, are obscured; and *delegation*, the passing on to others of the responsibility for carrying out specific actions. While our languages differ, the processes identified by Latour parallel closely many of the critical dimensions of "interlocking" and "distanciated" actor projects (see Long and van der Ploeg 1994 and 1995). Processes of mediation also lie at the heart of what I have termed social interfaces.

or encloses specific types of action. Instead, following Law (1994), we favor the idea of "ordering processes." The critical issue here is that "the structuring element is contained in the practices themselves: in the unfolding and, therefore, in that which is unfolded" (van der Ploeg 2003, 15). This is not, of course, to deny the impact of distant practices and arenas over which local actors may have little or no control (Long 2001, 62–65).

Various social and organizational processes function as a nexus of relations and representations where self-reflexive strategies mesh to produce a measure of accommodation between actors. These interconnections are crucial for understanding the articulation and management of actor interests and life-worlds, as well as for the resolution of conflicts. They constitute, that is, "new" or "reestablished" fields of enablement, constraints, and mutual sanctioning within which new embodiments of agency and social action take place. Actors' projects are realized in specific arenas and fields of action. Each project is articulated with other actors' projects, interests, and perspectives. This articulation might be regarded as strategic—consciously or not—in that the actors involved will attempt to anticipate the reactions and possible moves of the other actors and organizations. And the setting up of coalitions or modes of distantiation between particular actors is an intrinsic part of such action. For example, the various arenas in which farming interests are pursued contain what Benvenuti (1991) characterizes as "quasi-structures," for example, centrally regulated chains of commodity relations or particular networks of state organization that command authoritative allocative power. These forms or frameworks are not, however, disembodied entities; nor do they have a unilinear or uniform structuring effect on social practice or actors' choices. They link together around a common rationale or set of interests a number of participating social actors. The networks that result are composed of an extremely fluid set of properties that are the product of the interlocking and/or the distantiation of various actors' projects. At the same time, they constitute an important set of reference points and constraining/enabling possibilities that feed into the further elaboration, negotiation, and confrontation of actors' projects. Thus actors' projects and practices are not simply embedded in structural settings. Instead, it is through the complex ways in which they interlock that they create, reproduce, and transform particular patterned interactions. Thus, in the end, what we are dealing with is a series of interconnected social practices that provide some ordering (often contingently) to the networks of actors.

It is through the complex encounter and mediation of actors' projects that modes of ordering are generated. The emergence of such ordering processes is the outcome of the interplay of different self-reflexive strategies (Law 1994, 20), or what van der Ploeg and I have designated "interlocking actors' projects" (Long and van der Ploeg 1994). It is in this sense that we conceptualize structure. From the point of view of any one of the actors implicated in it, such a structure consists of a network of enabling and constraining entities (both human and "delegated" nonhuman, such as documents, machines, technology, and stocks of capital and material resources) and is internally heterogeneous (cf. Bijker and Law 1992, 300). That is, it is composed in multiple ways and looks and functions differently for actors situated, as it were, at different locations and adopting different stances within the social landscape.[13]

Such a conception of structure has certain analytical advantages. In the first place, it allows for the coexistence and interrelatedness of categories of actor that differ markedly in terms of their abilities to draw upon and extract benefits from particular resources, relations, and discourses. Second, it acknowledges the emergence—temporarily or more stably—of networks of power and of hegemonic tendencies that imply the predominance of certain values and ordering principles, however fragile. Third, it calls for a detailed appreciation of the coalitions, contingent relations, and shifts in meanings concurrently present within the overall network structure; and fourth, it transcends the notion of structure as a set of conditioning "externalities" by insisting that the pattern of interlocking actors' projects constitutes both a network of enablement and constraint as well as the embodiment and life source of agency and social action. It is my contention, therefore, that if we adopt this formulation we can swiftly move beyond the more structurally dependent interpretations of "resistance" and "hegemony," which I have critiqued in the earlier part of this chapter, to arrive at a more dynamic and nuanced understanding of processes of social struggle.

Counterdevelopment Discourse and Counterwork

In order to explore further the dynamics of agency, creativity, and strategic action in respect to struggles over "development"—and bearing in mind our

---

13. This idea of structure as consisting of contingent and heterogeneous elements tied together by a complex web of uneven linkages forged and remolded through social struggles was first formulated in the 1970s by myself and Bryan Roberts for analyzing patterns of regional development in the highlands of Peru (see Long and Roberts 1984). Later in the 1980s, van der Ploeg offered an analysis of the heterogeneity of agrarian structures through the specification of different "styles of farming" representing differential modes of ordering farm organization, the production process, and its development (van der Ploeg 1990).

special interest in issues of food and agriculture—I believe we need to give attention to processes of "counterdevelopment" and "counterwork" (Arce and Long 2000, 17–21, 8–9). The former depicts the ways in which mainstream development ideas and practices are challenged by alternative discourses emanating from actors directly or indirectly involved in the process of development intervention itself, or mounted by organizations or groups from outside. Such counterdevelopment discourses acquire a degree of legitimacy if they manage to capture succinctly some of the problems and contradictory tendencies of ongoing programs and can generalize from their experiences to make statements about "development in general." Counterwork involves multiple processes of reworking old modes of organization and meaning, and experimenting with new ones. In order to counter or redeploy successfully the authority and powers of existing state institutions and centrally organized development bodies, it is necessary to play not only with the inevitable ambiguities of policy texts and domains of implementation, but also with the ambivalences attached to various sets of social relations and value commitments (which become manifest in problematic situations where conflicts of loyalty arise), and finally to be bold enough to cross preconceived cultural frontiers (for instance, by associating with or confronting government officials of different status and class/political persuasions).

The following comments by Arturo Escobar crystallize the main points of our argument.

> Arce and Long (2000) have outlined a project of reclaiming and pluralizing modernity by focusing on the counter-work performed on development by local groups. The idea of "counter-work" highlights the ways in which the ideas and practices of modernity are appropriated and re-embedded in local life-worlds, resulting in multiple, local, or mutant modernities. . . . This notion includes a number of dimensions. First, a collective appropriation by social groups of the dominant practices of modernity into a shared cultural background that changes as a result.[14] This entails a dynamic repositioning of hegemonic elements within familiar contexts that

14. Escobar's footnote reads: "This first phenomenological feature is not explicit in Arce and Long, but it seems to me of paramount importance. In other words, phenomenologically speaking, any feature entering a human group from any where is by necessity actively re-absorbed into the local background of understanding—shifting such a background and, of course, the incoming element" (Escobar 2002, 36n7). See also Escobar (2003, 165 and 166n7) where he deploys the notion of "counter-work" to develop a new perspective on processes of population displacement in the Pacific area of Colombia.

involve the re-assembly of elements and the redrawing of boundaries, and a dis-embedding and re-embedding of Western standards within local representations of social life (including, of course, what counts as "modern"). Some of these processes were already visible in ethnographies of modernity. What the notion of counter-work adds is the idea of a very dynamic, always ongoing, endogenously generated task performed on incoming messages, elements, information, etc. that transforms both what we think of as the modern and the traditional. Such continuous processing is no longer a syncretism or hybridization of distinct cultural strands, but a series of self-organizing mutations driven by internal dynamics, even if often propelled by outside interventions. In the end counter-work may result in new power claims in terms of de-essentialising Western products of their superior power, or the empowering of a group's self-definition, even as it changes.[15]

If the Giddensian thesis suggests that globalization subsumes development—development becomes naturalized, even without its teleology and without assuming any single model, precisely because all people come to desire the same, namely commodities and markets—Arce and Long's conclusion is that every act of development is at least potentially an act of counter-development, and that every act of counter-development is potentially an alternative modernity—a "modernity from below." (Escobar 2002, 21)

Some Final Thoughts: How to Concretize These Conceptual and Theoretical Issues

All of this calls for a detailed understanding of the processes by which social meanings, images, identities, and social practices are shared, contested, negotiated, and sometimes imposed or rejected by the multiplicity of actors involved in specific situations. It also pushes us to inquire as to how far specific

15. Escobar's footnote reads: "In striving to explain this new conceptualization, Arce and Long rely on a series of metaphors that are also found in complexity theory (re-assembly, dynamic changes, self-organization, recursivity). I find some resonance between their notion of counter-work as an endogenous process and Maturana and Varela's notion of autopoiesis—change by drift with structural coupling to the environment and without loss of organization. . . . It is in part this language that in my view enables them to pluralize modernities without reintroducing through the back door a given meta-narrative" (Escobar 2002, 36n8).

actors' life-worlds, organizing practices, and cultural perceptions remain relatively autonomous of, or are "colonized" by, wider ideological, institutional, and power frames. Such dimensions, of course, are central to the types of organizational and value-related concerns that arise within contemporary agriculture and food systems. That is, we must ask what kinds of agency and political space farmers, input suppliers, food processors, traders, supermarket managers, R&D agriculture and food industry experts, and the wide variety of consumers and their organizations actually command. What interests and normative positions do they represent? And what leverage do they have over the policies and politics of agricultural change?

This question of how and to what extent they have leverage is, of course, complex. One should not assume that the supposedly dominant actors will win out, since there is ample evidence throughout the history of agricultural development that small-scale farmers and/or consumers can constitute a formidable force for change.

The same applies to the present-day antiglobalization lobby that has effectively toned down the neoliberal stance of the IMF, World Bank, and WTO in favor of supporting the economic plight of poor populations in the developing countries through a series of new poverty alleviation measures. The movement started out as a relatively small and incredibly heterogeneous mixture of protesters hell-bent on disrupting the meetings of these multilaterals and those of the G7 nations, but now it is much more organized and focused politically. The 2002 World Social Forum held in Porto Alegre—organized by a consortium of NGOs, academic groups, and sympathetic practitioners—hosted a series of hard-hitting debates that exposed the shortcomings of global neoliberal models and proposed alternatives that would be more equitable, just, and sustainable. One such solution was geared to the promotion of what delegates called an "Economy of Solidarity." Over the past few years, then, antiglobalization ideas have indeed reached a mass audience through the media and Internet, as well as through widely advertised public meetings. This concern for problems of globalization is matched by the rise of the Fair Trade movement and various environmentalist associations that are also setting agendas for "alternative development" that likewise reject crude market-led strategies.

Although it is tempting to conceptualize these kinds of social struggle in terms of the opposition of "local" versus "global" interests, or as protests "from below" aimed at hegemonic institutions and power holders, an actor-oriented perspective adopts a more catholic point of view. That is, it aims to elucidate the precise sets of interlocking relationships, actor "projects," and social practices that interpenetrate the various social, organizational, symbolic, and geopolitical spaces. In so doing, it focuses on the ongoing counterpoint between "sites of authority" and challenges "from below" and "above." Moreover, it does this without occupying the moral high ground of populism, or affirming the philanthropic or strategic goodwill of so-called well-intentioned "dominant" actors.

Social struggles of these kinds always center upon actor-defined issues, resource scarcities, and critical events; and it is precisely through a detailed study of these processes that one can reveal the contestation or intermeshing of differing discourses, interests, rewards, and practices.

REFERENCES

Arce, Alberto, and Long, Norman, eds. 2000. *Anthropology, Development, and Modernities: Exploring Discourses, Counter-tendencies, and Violence.* London: Routledge.

Archer, Margaret. 2000. *Being Human: The Problem of Agency.* Cambridge: Cambridge University Press.

Baumann, Gerd. 1996. *Contesting Culture: Discourses of Identity in Multi-Ethnic London.* Cambridge: Cambridge University Press.

Benvenuti, Bruno. 1991. "Towards the Formalisation of Professional Knowledge in Farming: Growing Problems in Agricultural Extension." Paper presented at the International Workshop on Knowledge Systems and the Role of Extension, Hohenheim, Germany.

Bijker, Wiebe E., and John Law, eds. 1992. *Shaping Technology, Building Society: Studies in Sociotechnical Change.* Cambridge: MIT Press.

Crehan, Kate. 1997. *The Fractured Community: Landscapes of Power and Gender in Rural Zambia.* Berkeley and Los Angeles: University of California Press.

Escobar, Arturo. 2002. "Borders, Networks, Difference: Social Movements and the Question of Modernity in Latin America." Paper presented at the Third International Congress of Latin Americanists in Europe, Amsterdam, 3–6 July 2002.

———. 2003. "Displacement, Development, and Modernity in the Colombian Pacific." *International Social Science Journal* 55 (March): 157–67.

Foucault, Michel. 1972. *Archaeology of Knowledge.* London: Tavistock.

———. 1981. *The History of Sexuality.* Vol. 1. Harmondsworth: Penguin Books.

Gibson, James J. 1979. *The Ecological Approach to Visual Perception.* London: Houghton Mifflin.

Gibson-Graham, J. K. 1996. *The End of Capitalism (as We Knew It): A Feminist Critique of Political Economy.* Oxford: Blackwell.

Giddens, Anthony. 1984. *The Constitution of Society: Outline of the Theory of Structuration.* Berkeley and Los Angeles: University of California Press.

Gramsci, Antonio. 1971. *Selections from the Prison Notebooks.* Ed. and trans. Quintin Hoare and Geoffrey Nowell Smith. New York: International Publishers.

Granovetter, Mark. 1985. "Economic Action and Social Structure: The Problem of Embeddedness." *American Journal of Sociology* 91 (3): 481–510.

Hale, Charles R. 1994. *Resistance and Contradiction: Miskitu Indians and the Nicaraguan State, 1894–1987.* Stanford: Stanford University Press.

Hutchby, Ian. 2001. *Conversation and Technology: From the Telephone to the Internet.* Cambridge, UK: Polity Press.

Joas, Hans. 1996. *The Creativity of Action.* Chicago: Chicago University Press.

Kearney, Michael. 1996. *Reconceptualizing the Peasantry: Anthropology in Global Perspective.* Boulder, Colo.: Westview Press.

Latour, Bruno. 1994. "On Technical Mediation—Philosophy, Sociology, Genealogy." *Common Knowledge* 3 (2): 29–64.

Law, John. 1994. *Organising Modernity.* Oxford: Blackwell.

Long, Norman, ed. 1989. *Encounters at the Interface: A Perspective on Social Discontinuities in Rural Development.* Wageningen Studies in Sociology No. 27. Wageningen, The Netherlands: Wageningen University.

———. 2001. *Development Sociology: Actor Perspectives.* London: Routledge.

Long, Norman, and Ann Long, eds. 1992. *Battlefields of Knowledge: The Interlocking of Theory and Practice in Social Research and Development.* London: Routledge.

Long, Norman, and Bryan Roberts. 1984. *Miners, Peasants, and Entrepreneurs: Regional Development in the Central Highlands of Peru.* Cambridge: Cambridge University Press.

Long, Norman, and Jan Douwe van der Ploeg. 1989. "Demythologizing Planned Intervention: An Actor Perspective." *Sociologia Ruralis* 29 (3–4): 226–49.

———. 1994. "Heterogeneity, Actor and Structure: Towards a Reconstitution of the Concept of Structure." In *Rethinking Social Development: Theory, Research, and Practice,* ed. D. Booth, 62–89. Harlow, UK: Longman Scientific and Technical.

———. 1995. "Reflections on Agency, Ordering the Future, and Planning." In *In Search of the Middle Ground: Essays on the Sociology of Planned Development,* ed. Georg E. Frerks and Jan H. B. den Ouden, 37–59. Wageningen, The Netherlands: Wageningen Agricultural University.

Sayer, Derek. 1994. "Everyday Forms of State Formation: Some Dissident Remarks on 'Hegemony.'" In *Everyday Forms of State Formation: Revolution and the Negotiation of Rule in Modern Mexico,* ed. G. M. Joseph and D. Nugent, 367–77. Durham: Duke University Press.

Scott, James C. 1985. *Weapons of the Weak: Everyday Forms of Peasant Resistance.* New Haven: Yale University Press.

Smith, Gavin. 2003. "Hegemony." In *The Blackwell Companion to the Anthropology of Politics,* ed. D. Nugent and J. Vincent, 216–30. Oxford: Blackwell.

Turner, Victor W. 1974. *Dramas, Fields, and Metaphors: Symbolic Action in Human Society.* Ithaca: Cornell University Press.

Van der Ploeg, Jan Douwe. 1990. *Labor, Markets, and Agricultural Production.* Boulder, Colo.: Westview Press.

———. 2003. *The Virtual Farmer: Past, Present, and Future of the Dutch Peasantry.* Assen The Netherlands: Royal van Gorcum.

Villarreal, Magdalena. 2001. "Engineering the Future for Women-Subject-to Development." Paper presented at Workshop on Agency, Knowledge, and Power: New Directions in the Sociology of Development, Wageningen University, The Netherlands, 14–15 December.

Willis, Paul. 1977. *Learning to Labour: How Working-Class Kids Get Working-Class Jobs.* London: Saxon House.

INSTRUCTOR'S RESOURCES

Key Concepts and Terms:
1. Agency
2. Counterwork
3. Discourse
4. Resistance
5. Structure

Discussion Questions:
1. Discuss the ways in which Long's notion of agency differs from the formulation advanced by Friedland.
2. What are the advantages of an actor-oriented perspective, according to Long? What are its drawbacks?
3. In what ways are the concepts of agency, counterwork, and resistance useful in studying change initiatives in the agrifood system?

Agriculture, Food, and Environment on the Internet:
1. WWW Virtual Library Sociology: http://socserv.mcmaster.ca/w3virtsoclib.
2. Sociological Theories and Perspectives: www.sociosite.net/topics/theory/php.

Additional Readings:
1. Archer, Margaret. 2000. *Being Human: The Problem of Agency.* Cambridge: Cambridge University Press.
2. Long, Norman. 2001. *Development Sociology: Actor Perspectives.* London: Routledge.

PART II

MAKING ROOM FOR AGENCY

4

COUNTERHEGEMONY OR BOURGEOIS PIGGERY?
FOOD POLITICS AND THE CASE OF FOODSHARE

Josée Johnston

Back in 1985, before there was an entire television channel devoted to food, before the omnipresence of Williams-Sonoma and a twenty-four-hour food network, and before *Vogue* declared food "the new sex,"[1] Barbara Ehrenreich identified the growing cultural importance of food: "Among the upscale, trend-setting people who are held up for our admiration in commercials for credit cards and wine coolers, *food appears to be more fascinating than either sex or trivia games*" (1985, 18, emphasis added).

In a three-page essay, Ehrenreich didn't have much space to elaborate on how this had happened, but she did suggest that interest in healthy food, organic oatmeal, and non–genetically modified salad greens is a new variant on an old theme of social status and conspicuous consumption (see Bourdieu 1984; Veblen 1899/1994). While Ehrenreich's essay took a jocular tone, the stratification of our food system is no laughing matter, particularly when it is linked to serious health risks for the poor (Critser 2003; Kempson et al. 2002; Sobal and Stunkard 1989; Drake 1992).[2] Post-Fordist niche markets for organics and "natural" foods capitalize on the food anxieties and disposable income of highly privileged economic strata, while diluting organic standards and siphoning resources from southern producers through global food chains (Buck, Getz, and Guthman 1997; Guthman 1998; Friedland 1994). A stratified food system perpetuates class inequality on a global scale, yet public

---

1. The British edition of *Vogue* made this assertion in 1996. See N. Foulks, "Eat Me," *Vogue,* UK ed., April, 40–43.

2. The relationship between diet and class is complex. While low-income groups do appear to eat a diet that is less nutritious, upper-income groups also suffer from poor eating habits (see Crotty 1999). Epidemiological evidence suggests that the poor health of low-income groups cannot be reduced to diet, and that there are multiple determinants of ill health (Crotty 1999; McIntyre 2003).

attention seems to be diverted elsewhere—to the health benefits of flaxseed oil, protein-based diets, and low-carb breads. As Ehrenreich writes, "Who remembers the working class now? Once we would have picketed any venue that dared to sell scab grapes or lettuce. Now we are mollified by the posters at the Whole Foods market, introducing its many personable employees" (1994, 5; see also Kauffman 1991).

Food is clearly a marker of social class and a form of cultural capital that elites use to consolidate and reproduce social inequality (see Levenstein 1993; Mennell 1986; Bourdieu 1984). But food is more than a simple instrument of class oppression. Food also represents an entry point for political engagement, making it problematic to single-handedly categorize all interest in food as bourgeois piggery. While hungry people are not always revolutionaries, food has historically played a role crystallizing counterhegemonic resistance to unjust social relationships. In the 1960s the counterculture reacted against the technological quick fixes of processed cheese and Wonder Bread, rejecting "white" food as elitist, and declaring solidarity with the world's "brown" people by eating brown rice, brown bread, and brown sugar (Belasco 1993, 48–50). Even earlier, food movements in the 1830s in the United States embodied a reaction against developing commercialism and its effects on food habits and moral behavior (Gusfield 1992). Food plays a similar catalytic role today, providing a tangible reference point for the most nefarious elements of neoliberal globalism: the corporate control that threatens regional food cultures, the inequality that perpetuates food insecurity for millions, and technological food fixes like irradiation and genetic manipulation. While the corporate food complex aspires to create "dependency to ensure corporate profits" (Kneen 1999, 161), alternatives to the corporate agrifood industrial complex are on the rise, as detailed in other chapters in this volume (see also Kloppenburg et al. 2000; Hinrichs 2000). Today's consumers can choose Fair Trade coffee and bananas, drink milk produced in an organic cooperative, or grow vegetables in a community garden. While some alternatives are profit-oriented, other projects contain explicit anticapitalist agendas and purposely set out to define food as a fundamental human right—not a mere commodity.

Despite reasonable skepticism on the left about culinary obsessions in North America, I argue that we cannot simply reject the renewed interest in food as bourgeois piggery—even though the risk of capitalist co-optation is present in all counterhegemonic projects (Campell 2001, 354). It is impossible to create simple categories of "revolutionary" and "co-opted." The challenge for food activists and scholars is to identify subtle degrees of emancipation and domination in food politics. The first step along this path is to clarify what it means to have agency in the global food system—a question that, I argue, requires an understanding of power.

It is difficult to conceptualize agency without considering human agents' dialectical relationship with capitalist structures and cultural hegemony. The Marxist tradition stemming from Hegel suggests that capitalist processes contain their own internal contradictions, and nineteenth-century writings from Antonio Gramsci (1971) and Paulo Freire (1970/2000), as well as insights from antiglobalization struggles (Conway 2004), suggest that emancipatory actions stemming from capitalist contradictions are necessarily pluralistic. Put differently, there is no one political actor (or vanguard) that can overthrow capitalism, or even a more specific target like the agrifood industrial complex. To appreciate a pluralistic conception of agency, we must understand a bit about where power is located, and how it operates.

Insights from French philosopher Michel Foucault and poststructuralist theory emphasize that power is not simply centralized in a mythical capitalist headquarters where regulations are enforced from the top down but flows in a capillary, diffuse fashion. According to Foucault, the power to govern stems less from top-down regulation than from bottom-up normalization, through which people become their own "jailers," participating in regimes of self-punishment, self-monitoring, and normalization (1990, 1977). If power is exercised in multiple locations (and not simply from a centralized power holder), then resistance requires multiple points of contact, as well as multiple projects that seek to problematize, or "de-normalize," the exploitive relationships we have grown accustomed to in consumer-capitalist societies. One example is the conviction, "normal" for most North Americans, that food should be available at a bargain price, a belief that relies on labor exploitation and environmental exhaustion at multiple points along the commodity chain. (See the following chapter, by Aimee Shreck, on how this happens through coffee commodity chains.) A further implication flowing from a capillary conceptualization of power is that individual actions *do* have important social consequences, an argument that feminists make when they say that "the personal is political," even if the effects of individual actions are not easy to disentangle from their social contexts.

Foucauldian insights about the diffuse, fragmented nature of power in late capitalist societies have been exceptionally important and have added greater subtlety and nuance to our understanding of agency as necessarily multifaceted. But they have also led to the troubling tendency to fetishize (or romanticize) resistance so that agency, power, and responsibility are identified everywhere and anywhere. A more balanced approach to understanding power recognizes both sides of the coin: that power is diffuse and fragmented, resting not just with heads of state but with the agency of multiple actors from "below," but that power is *simultaneously* centralized at the "top" through key actors like agrifood corporations, regulatory bodies like the World Trade Organization, and the geopolitical and economic power of the United States and other core states in the world system.

This balanced approach recognizes that power is not simply a zero-sum game in which "rulers" govern their hapless minions, but neither does it see power as an infinite fragmentation where no person or agency holds responsibility. Instead, agents are thought to have power (and agency) when they possess *the capacity to affect outcomes* (see Schaap 2000, 123). We recognize that certain actors and institutions bear responsibility for domination—for constraining the power of others to shape their own outcomes—but we must also identify the positive, collective side of power, which is *empowerment.* Empowerment is necessarily a group project; it requires the cooperation of many people and collectivities to generate a meaningful capacity to affect outcome; it is a concept that moves us away from naïve, individualistic conceptions of agency. While Foucault made clear that norms and behaviors are used to discipline individuals and communities, a neo-Marxist reading of Foucault emphasizes that these norms and behaviors are created by collectivities in a capitalist world system. This makes it important to identify and assign responsibility to agents who perpetuate norms and behaviors that normalize exploitation and constrain empowerment. All of this leads us to a very important final question about power and agency: what kind of norms and behaviors are critical to promoting greater sustainability and justice in the agri-industrial food system? Put differently, what kind of norms and behaviors can be considered counterhegemonic in the sense that they expose the exploitive norms and behaviors that normalize oppression and create new norms of empowerment?

To answer this question, I develop a framework for establishing counterhegemonic criteria for transformative food politics that rests on two key normative goals: (1) reclaiming the commons, and (2) creating postconsumer values. The utility of these counterhegemonic principles is best understood not through abstract philosophical debate but through the study of social movement struggles. To that end, I examine FoodShare—one of the most visible and successful community food security agents in Canadian food politics. While one case study cannot encompass the complexity of food politics or agency in the food system, or single-handedly transform the agrifood system, success stories like FoodShare speak to the challenges and obstacles facing those who seek to determine different outcomes in the food system—more sustainable, just, and equitable outcomes. The FoodShare story demonstrates a form of empowerment through its capacity to positively affect outcomes for participating community members. At the same time, the greater power of the agri-industrial food system limits the success of this community food security project, constricting both its agency and its capacity to promote more substantial counterhegemonic values and behaviors.

Establishing Counterhegemonic, "Anti-Piggery" Criteria

My interest in developing criteria for counterhegemony grew out of frustration with labels like "sellout," "bourgeois," "co-option," and "NGO-ization." When they use these terms, nonprofit organizations are criticized for cushioning the inequities of corporate capitalism and allowing the welfare state to dismantle itself, while diverting attention from pressing social and ecological issues (see Seccombe 2002). A version of this critique focuses on NGOs and community food security projects specifically, which are criticized as a patchwork approach to food provisioning that does little to address the reform of the welfare state or structural issues of poverty and inequality (Tarasuk and Davis 1996, 74; McIntyre 2003, 51). While activists busy themselves with food drives, heirloom tomatoes, and garden plots, the state surreptitiously diminishes its responsibility for basic needs like food and shelter. American Marxist Joel Kovel agrees that "a lovely garden is a wondrous thing" but cautions that "given the current predicament, [this type of social project] is a signpost and not an end" (2002, 156). Put differently, community gardens and school lunches can be life-affirming and hopeful, but for some critics they are comparable to rearranging deck chairs on the *Titanic*.

But is it fair to reduce social experiments like community gardens or farmers' markets to "signposts" rather than legitimate counterhegemonic "ends"? Clearly, tough questions need to be asked. Should poor people have access to food staples through community gardens and collective kitchens, while middle-class folks garden as a hobby and shop at upscale supermarkets? Are artisanal organic products luxury items out of reach to the poor, who are condemned to eat mass-produced industrial food? Do community food security projects exist on a scale that can challenge the global scale of agri-industrial food chains? (See Johnston and Baker 2005). These types of civic agriculture projects are not a magic bullet, but it seems simplistic to discount them without considering the complexity of the social change process, the urgent need for innovative approaches to social and ecological problems (see Homer-Dixon 2001), and the possibilities for empowerment they offer.

So how should we frame food politics, and what criteria could be used for both affirmation and critique? While food is a material necessity, its symbolic value makes it a highly complex social phenomena, capable of being used to both reinforce and erode relations of domination? A Gramscian concept of hegemony helps negotiate such complexity, since it avoids determinist, top-down accounts of social change and recognizes the importance of cultural politics in struggles for social justice. Carroll (1992, 8) defines hegemony as "a historically specific *organization of* consent that rests upon—but cannot be read off—a practical material base." The concept of hegemony emphasizes that elites cannot rule by force alone—cultural leadership is required to achieve cultural consent, which reinforces class inequality and often works to "short-circuit attempts at critical thinking" by working its way into common sense (Smith 2001, 39). While we recognize that power is concentrated in key capitalist agents and organizations (the "top" of the global system), neo-Gramsican approaches also see power and agency at the "bottom," resting in the hands of citizens, nonprofit organizations, and social movements.

Because the power of the "top" and "bottom" interact dialectically, hegemony is best thought of as a moving target rather than as a fixed set of ideas for mass rule (Gramsci 1971, 12, 276). For hegemony's cultural leadership to be effective it must change along with social and historical circumstances; cultural consent is never permanent but is always in the process of being contested (Hall 2001, 97; Roseberry 1994, 360–61). Hegemony's continual evolution means that counterhegemonic strategies are also historically specific, contextual, and continually evolving. While classic Marxist theory focused on labor exploitation, contemporary theorists and activists better understand the myriad forms of gender, race, and ecological exploitation in capitalist societies. In her account of environmental politics in Canada, Adkin broadly defines counterhegemonic challengers as "critical of capitalist accumulation, of productivism, of science as domination of nature, of the prevailing ideologies of science and technocracy, of relations of subordination-domination (gender, racial, heterosexual), and of the institutions and social practices that underpin such relations" (1992, 136).

As this definition suggests, one key element of contemporary hegemony is the belief in perpetual economic expansion—a notion increasingly challenged as pollution, resource depletion, and climate change become harder to ignore (Smith 1998, 5; McMurtry 1998). While mainstream economists see ecological degradation as an inconvenient "externality" of economic growth, others suggest that the problem is more fundamental, rooted in basic contradictions of capitalist markets that threaten the long-term viability of human life on the planet (O'Connor 1996; van der Pijl 1998; Kovel 2002). Despite the growth of public awareness of environmental issues and the rise of green marketing, current environmental solutions do not appear to be radical enough to stem the tide of ecological exhaustion and species extinction taking place.[3] This makes it imperative that we integrate ecology into theories of hegemony, power, and agency. In this vein I suggest that the two ecosocial criteria mentioned above—reclaiming the commons and creating postconsumer needs and desires—are particularly important for a counterhegemonic food politics based on empowerment and ecological integrity.

Reclaiming the Commons

While the language of sustainability has become ubiquitous in business circles, "reclaiming the commons" is a phrase increasingly heard in the global justice movement (Johnston 2003; Goldman 1998, 14). This phrase reflects a general interest in reorienting economies away from an exclusive focus on commodification and maximizing profit, and toward a more equitable and sustainable provisioning of human needs. While the commons have been defined in many ways, ecophilosopher John McMurtry usefully describes them as "human agency in personal, collective or institutional form which protects and enables the access of all members of a community to basic life goods" (1999, 204).[4] Put differently, the civil commons represent what people do as a society to "protect and further life, as distinct from money aggregates" (McMurtry 1998, 24). Mano similarly describes an "economy of care and concern"—a noncommodified realm where needs are met not exclusively

---

3. An article in *Nature* documented a "coherent pattern of ecological change across systems" resulting from already observable signs of global warming, and predicts an impending avalanche of species extinction (Walther et al. 2002). E. O. Wilson (1999) speaks of a sixth spasm of extinction, which, unlike the previous five extinction waves that occurred over the past 500 million years, is caused primarily by human beings.

4. McMurtry distinguishes a life-good from a commodity using two criteria: (1) freedom from a price barrier (while markets can be used to distribute life-goods, they cannot be restricted to those with resources), and (2) the property of enabling vital life capabilities, which includes not just the capacity to be physically alive but the broad human range of thinking, acting, and feeling (2001, 827, 837).

through commodities but through a combination of need reduction, need prevention, cooperation, and collective approaches (2002, 69). The commons are not a utopian exercise but a mode of human organization that exists in varying degrees within traditions such as collective grazing areas, universal healthcare programs, public parks, and community kitchens. A commons approach to transportation, for example, rejects an exclusive reliance on individual cars and instead meets human needs for exercise, clean air, sociality, and transportation through publicly supported transit solutions combined with well-planned communities to facilitate walking and bicycling.

While a healthy economy would meet human needs through a variety of approaches, the predominant mode in advanced capitalist societies is through individual commodity production and consumption, a trend that is self-reinforcing and undermines collective, noncommodified methods of meeting human needs (Mano 2002, 70, 82). As antiglobalization activists have noted, everything in the global economy is potentially up for sale: plants, genetic codes, traditional knowledge, sexuality, children, and human organs. With commodification comes heightened dependency on the market for meeting basic needs. In his study of the transnational agricultural corporations Monsanto and Cargill, Kneen finds that an "intentional by-product of [corporate] strategies is the elimination of self-reliance, self-provisioning, and autonomy" (1999, 163).

The hyperdevelopment of commodities on a global scale can be understood as an enclosure of the commons, a development that threatens self-reliance and the cooperative spirit found in many local ways of life. This has led activists worldwide to call for public policy interventions and local organizing efforts to "create a parallel economy of care and connection that can counter the negative effects of the domination caused by the economy of commoditization" (Mano 2002, 99). In this context, food politics can be considered counterhegemonic when it resists the enclosure of the commons and works to ensure that access to basic life-goods like food can be met through noncommodity channels, particularly when sufficient purchasing power is lacking. Counterhegemonic projects reclaim the commons by insisting that "our world is not for sale," demanding democratic control over food production and consumption, and resisting the idea that these needs are "commodities" best controlled through commodity markets and large corporations (Bove and Dufour 2001). Reclaiming the commons does not necessarily mean that markets and individual consumption styles are eradicated, but it does demand that markets be reembedded in social structures that ensure that nutritious, sustainable food goes not only to those who can afford it but to everyone, and that alternate modes of provisioning—through cooperative provisioning—are equally developed.[5]

Not only is commodification the dominant mode of meeting food needs in capitalist economies, but it has also been characterized by a dramatic separation between food production and consumption, along with a concentration of power and control in agribusiness monopolies that affords minimal agency for average citizens (Duncan 1996; Winson 1993). Harvey describes how globalization "brings together different worlds (of commodities) in the same space and time," and "conceal[s] almost perfectly any trace of origin, of the labour processes that produced them, or of the social relations implicated in their production" (1989, 300). This extreme distancing opens the door to risky and disempowered ecological behavior, as "the viability of society becomes both less transparent and more precarious" (Duncan 1996, 39). For people with a close, direct relationship to the land, destroying natural resources like well water or arable soil is like shooting yourself in the foot. When you have an intimate connection to the natural commons that sustain you, you are more likely to see the consequences of degradation and to act accordingly. Under globalized capitalism, however, the consumers of Chilean grapes or California strawberries are profoundly disconnected from the conditions of production. Not only does this allow elite consumers to live off the carrying capacity of distant others, it encourages a profound anthropocentrism, as people lose any sense of how they depend on the natural world to meet basic needs.

In short, if counterhegemonic food practices are to reclaim the commons, they must restrict commodification, limit the separation of food production and consumption, access life-goods through cooperation and communal approaches, and rethink the categories of "needs" and "wants." This means building accessible connections between farm folk and city folk and creating spheres where people question their food preferences and even learn to feed themselves—either through their own labor or through the labor of people with whom they share bonds of citizenship obligations and responsibilities. This is not a utopian project that can be achieved through one policy measure or food project. It is a long-term, incremental process of reclaiming social life from an excessive reliance on market logic and global

5. Complex debates about the role of markets in modern capitalism cannot be explored here. For a discussion of the historical importance of markets in capitalism, see Wood (1999); on the connections between markets and ecological exhaustion, see van der Pijl (1998). A more positive assessment of markets and the environment is found in Hawken, Lovins, and Lovins (2000), but Guthman assesses market dynamics in organic agriculture and reaches more pessimistic conclusions (2002, 305–6).

commodity chains. Theoretically, we can understand these types of reclamation projects as dialectical critiques of commodification and capitalist enclosure. More concretely, we can understand urban farmers, community kitchens, rooftop gardens, and community-supported agriculture as engendering collective empowerment and personal agency that is based not just on how much you can buy but on values of sufficiency, equity, and stewardship of the commons. This introduces the second criterion for a counterhegemonic food politics: creating desires beyond the realm of market-driven commodity production.

Creating Postconsumer Needs and Pleasures

Food politics can also be counterhegemonic when actors resist the ideology of consumerism and create postconsumer needs, desires, and pleasures that are ecologically sustainable. Consumerism can be broadly defined as the "set of beliefs and practices that persuades people that consumption far beyond the satisfaction of physical needs is, literally, at the centre of meaningful existence and that the best organized societies are those that place consumer satisfaction at the centre of all their major institutions" (Sklair 2001, 5). Consumer societies valorize an ideal of consumer sovereignty in which individual choice is valued more highly than collective action to combat social problems, and the ideals of citizenship—a concept embodying rights *and responsibilities* for the common good—are minimized. The consumer sovereignty ideal presents a romanticized view of individual consumer agency operating in a socioecological bubble, where one's actions are disconnected from underlying ecological variables. The pursuit of individual self-interest through consumption is thought to automatically produce a greater common good, and commodities in the marketplace are presented as a privileged medium for negotiating individual identities and needs (Slater 1997, 28–29; Keat 1999).

The ideal of consumer sovereignty involves strong popular and emotional attachments, particularly since it appears as one of the few arenas of individual agency in modern life (Slater 1997, 27). Consumers at the mall, cash in hand, are widely considered "empowered"—able to affect the outcome of their own individual shopping decisions, even if they cannot determine whether these choices are sustainable and fairly produced, or whether their budgets will cover all of their needs and wants. Consumerism, with its atrophied conception of individual agency, is part of hegemonic common sense.

Postconsumer projects, which critique consumerism as soul-wrenching, socially isolating, and ecologically devastating, and present collective ideals of empowerment, are confined to minority elements of countercultures. There is some evidence that this is beginning to change, as the risks of over-consumption register in the public sphere—a thinning ozone layer, landfills piling up, pesticide use linked to cancer rates, visible connections between industrial food production and public health problems—and the costs of northern consumption habits come under increasing scrutiny (Sachs 1998). Alternative consumer projects like Fair Trade have been formed to push for progressive social change through collective consumption projects (see the following chapter in this volume by Shreck, as well as Johnston 2002; Shreck 2002). Awareness of consumer dystopia is growing even among those who do not want to alter their shopping habits.

But is fear of ecological Armageddon enough to mobilize counterhege-monic projects against consumerism? One key lesson of new social move-ments is that sustained mobilization cannot be based exclusively on fear or repression but must create a foundation of shared identities and understand-ings of the good life. As Carroll writes, "if social movements can thus be viewed as counter-hegemonic in the sense of *opposing* the existing order, there remains the vexed question of whether and how counter-hegemonic politics can be defined in a more proactive, visionary sense" (1992, 10).

Reasoned argument is not always enough to sustain social action, even on the most pressing ecological or social issues; positive, pleasurable alterna-tives to consumer seductions are required. Few have articulated this idea more eloquently than ecophilosopher Kate Soper, who argues for a vision of pleasures and desires not based on human or ecological exploitation (1999, 1990). According to Soper, attempts at political mobilization cannot simply emphasize what one should *not* enjoy as a wealthy consumer but must also involve a vision of the joys and pleasures of a postconsumer life—a way of living that is not primarily tied to identity construction through com-modity purchase. Soper does not claim that substitutes for "market-driven life-styles" are fully developed, and writes that "no one could deny that there are very few signs at the present time of any imminent shift away from these growth-oriented and self-regarding political commitments" (1999, 129, 119). Yet the case is not closed. Soper suggests that we are witnessing the emergence of a "new more contradictory structure of consumer needs," one in which "consumers are looking to alternative life-styles in order to escape the unpleasurable by-products of their own formerly less questioned sources of gratification" (120). This search is not simply based on a wish to eliminate exploitation but is also rooted in a desire to escape consumerism's by-products (e.g., personal exhaustion, stress, credit card debt, and environmental destruction). As Soper points out, there is a growing and palpable sense of alarm about ecological exhaustion and global disparities of wealth, as consumers become "uneasy about the ways in which the pursuit of first world affluence protracts and exacerbates deprivation elsewhere" (119).

While not everyone feels the negative effects of overconsumption, awareness of them is becoming more mainstream all the time. As Giddens writes, "the risks of ecological catastrophe form an inevitable part of our horizon of day-to-day life" (1991, 4). For example, Canadian newspaper columnist Heather Mallick wrote a 2002 pre-Christmas column in the Toronto *Globe and Mail* detailing her list of consumer pleasures from the previous Saturday—"beaded velvet jewellery boxes, a shearling overcoat, the collected works of David Sedaris, 45 pairs of socks, a facial at Estée Lauder, two Christmas wreaths, a purse by Longchamps and a hair crimper"—along with numerous taxi rides, various meals, and electronic devices (30 November 2002, F2). Mallick expressed guilt about these purchases, recognizing that "Mother Earth is sagging beneath the weight of our demands," and then shockingly concluded that given her incredible burden on the planet, the ecologists are right: she would probably be better off dead and composted, buried in a flannel nightgown and providing bone meal for plants.

While Mallick cheekily presents death as the only realistic alternative to hyperconsumption, there are other, less nihilistic alternatives outside the restricted world of upper-middle-class angst. The point is not that we would be better off dead but that counterhegemonic projects must speak to the human need for pleasure, beauty, and desire. Counterhegemony cannot simply deny consumerism's pleasures but must work with the anxiety produced by excessive consumption and develop attractive visions of alternative ways of life, where empowerment is sought at places other than the shopping mall.

But is it possible to reconcile Soper's argument about the need for post-consumer pleasures with Ehrenreich's concern about the co-optation of the Left by bourgeois piggery? The two arguments are often presented as mutually exclusive, separated by different groupings of scholars and disciplinary specializations (some study food's pleasures, while others focus on food's injustices). A richer field of food studies would productively engage both these realms, bridging obstinate divides between culture and political economy, recipes and catastrophes, consumption and production. While market forces can transform food pleasures into exploitive profit opportunities, this does not always happen, nor does it occur in a straightforward way. Concerns

about lifestyle, daily meals, and physical health can be incorporated and commodified into hegemonic processes (see Guthman 2002), but they can also
work as a kind of gateway drug to larger politicization processes, expanding
reflexivity about the social and ecological costs of industrial food production (see DuPuis 2000; Giddens 1991, 222).

I have argued that hegemony is a useful conceptual tool for examining
such complexity because it describes a process of negotiation that can involve
genuine compromises by dominant classes. Counterhegemonic challengers
question the benefits of commodity markets, challenge the centralization of
power in a corporate-driven food system, and demand public policy solutions
and community control over food. Certain challenges may even be incorporated into common sense positions on food safety, nutrition, and cuisine.
This complexity compels the analyst to look for points of contradiction and
change (rather than search for ideologically "pure" agents of domination or
resistance), and suggests the need for empirical analysis of practical strategies
and struggles, often on a case-by-case, crop-by-crop, or sector-by-sector
basis. The counterhegemonic criteria summarized in Table 4.1 are intended
as starting points that emerge from critical praxis and will next be used to
discuss one such practical strategy—the good food box.

FoodShare and the Good Food Box (GFB)

The good food box is a product of a Toronto-based nonprofit called Food-
Share, an organization started in 1985 by Art Eggleton, who was mayor at the
time.[6] Eggleton's goal was to create a temporary relief agency to coordinate

Table 4.1   Counterhegemonic Criteria

1. Reclaiming the commons, a realm of social life that:	2. Creating postconsumer needs and pleasures that:
• Restricts commodification • Develops alternative modes of meeting life-goods (e.g., cooperation) • Decreases distance between production and consumption	• Challenge consumerism's hegemony • Provide a proactive vision, creating alternative pleasures and empowerment not based on ecological and social exploitation

6. Data on FoodShare were collected during participant observation and interviews carried out
between October 2002 and May 2003.

access to the growing number of food banks and create a humanitarian image to help his re-election bid. Eighteen years later, FoodShare has evolved from a shoestring operation into a well-known nonprofit with at least twenty staff members, hundreds of volunteers, an annual budget of more than $3 million, and a reputation for innovation. While FoodShare's origins lie in the charity model of food distribution, since the early 1990s the organization and its leaders have moved decisively toward *community food security* responses to hunger, explicitly endorsing principles of bioregional food production, community development, and social justice.

FoodShare's activities are divided into two offices.[7] The downtown Queen Street office organizes education, research, and community development. Programming includes community gardens, online education, baby food–making classes, salad bars in public schools, and a "Field-to-Table" education campaign and summer festival. There is also direct-mail fund-raising and a volunteer-run phone line that directs callers to food projects around the city. Debbie Field is the executive director of FoodShare and serves as the charismatic spokeswoman for the organization. While Field has an extensive background in labor politics and Marxist theory, she employs a flexible alliance strategy that seeks dialogue and partnerships with business partners, several of whom serve on the board of directors and secure corporate sponsorship for events.

The Field-to-Table warehouse is the second FoodShare office and is located in the east end of downtown Toronto. Projects at this location are production oriented and involve youth training and community development. The warehouse director at the time of my research, Mary-Lou Morgan, has a background in organic retailing. The Field-to-Table warehouse also houses an incubator kitchen (where women can begin small enterprises) and a catering operation that serves private clients and also makes "power soups" for soup kitchens and drop-in centers. Field-to-Table is the only certified organic operation within the city limits. An impressive greenhouse operation grows sprouts and seedlings hydroponically for a market garden at a nearby center for addiction and mental health. The warehouse also contains a substantial composting operation, beehives that produce organic honey, and a "Focus on Food" training program where "youth at risk" (e.g., street kids and recent immigrants) receive basic employment and lifeskills training while working in the warehouse, greenhouse, and kitchen.

---

7. The physical organization of FoodShare has changed since this research was conducted in 2003. While the programming remains similar, in October 2006 FoodShare consolidated its various operations in one physical space in the Dufferin Grove neighborhood of central Toronto.

The core of these activities, and the linchpin of FoodShare more generally, is the good food box program (GFB). The GFB emerged from an afterwork discussion between members of the Toronto Food Policy Council,[8] when a farmer and a community development worker puzzled over how to connect a crop of small and unmarketable potatoes with low-income consumers. A feasibility study funded by the council gave birth to the "Field-to-Table" concept. The initial idea was to use a truck to deliver local produce to selected locations (e.g., low-income apartment complexes and seniors' residences), but this evolved into a weekly box of fresh produce delivered by local coordinators. The GFB began as a one-size-fits-all model but now includes a smaller box, a fruit basket, a "reach-for-five" box aimed at seniors and those with limited mobility, a small and large certified organic box, and a wellness box delivered to women being treated for breast cancer.

Producing the GFB is a complex and time-consuming process. Each month a dedicated core of staff and volunteers put in hundreds of hours to produce three to four thousand boxes, affecting the diets of up to eight thousand Torontonians (assuming that each box feeds two or three people). The cost of the produce is paid by the customer, but the infrastructural support is financed by FoodShare, which is funded by public and foundation funding, a mass-mail donation base, and revenue from its product sales. (It is a registered nonprofit organization, so donors can deduct contributions from their taxes.)

Determining the efficacy of the GFB as a counterhegemonic tool requires that we first understand how it differs from a for-profit delivery service. First, the GFB is not organized to deliver maximum profits but is organized around the belief that good food should be affordable, accessible, and based on sustainable production methods. The organization cannot afford to lose money, but profits are secondary to these primary goals. Subsidization by public funds and private donations allows staff to purchase fruits and vegetables for the GFB using extramarket criteria—especially the commitment to buy locally and thus support local farmers. How does this happen? Food is bought through the Ontario Food Terminal, but in the growing season it is also purchased directly from local farmers with whom the organization has established a relationship over the years. FoodShare commits to paying a fair price for the produce, not just the cheapest price it can get, and supports small-scale producers through crop failures and processing mishaps.

8. The Toronto Food Policy Council was formed in 1991 as a subcommittee of Toronto's Board of Health. Members include city councilors and volunteer representatives from consumer, business, farm, labor, multicultural, antihunger, faith, and community development groups.

Using an incubator model of entrepreneurship that helps small business grow in a supportive environment, FoodShare helps organic growers become established so that they can move on to service bigger clients and in greater numbers.

The other factor that differentiates the GFB from a for-profit delivery service is its commitment to delivering quality produce to low-income people. This does not happen in a perfect or absolute way, but it does happen to a far greater degree than would be possible with a for-profit service. FoodShare has been criticized for not reaching the poorest of the poor in Toronto, and for servicing an exclusively middle-class group with its pricier organic box. While a small subset of the customer base (7 percent) has household incomes of more than $70,000 (Canadian dollars), at least half of the regular food boxes go to families that fall below the poverty line. A substantial number of boxes go to mainly low-income volunteers who help pack the box (and who have to be put on a rotational waiting list, given the high demand). FoodShare argues that the multiclass makeup of GFB customers broadens its appeal and presents a way for low-income patrons to access affordable fruits and vegetables without stigma.

So how does the GFB fare when considered against the two criteria for counterhegemonic food politics? The first criterion involves reclaiming food from the capitalist commodification drive, improving access to healthy food through cooperative channels, and lessening the distance between production and consumption. As noted above, the GFB's goal is not to maximize profits but to reclaim healthy food as a human right and an ecological necessity. The program is centered on the belief that fresh produce is a basic lifegood, not a commodity that only privileged classes should enjoy, and this means that the box reaches a relatively large percentage of low-income clients. While the GFB prioritizes low-income consumers, it is deemed more important to support local farmers than to aggressively pursue the lowest possible price for food. Satisfying the competing demands of social justice and ecology is a delicate balancing act that transcends conventional business organization; it has led FoodShare to develop an innovative third-sector mode of operation that uses markets but relies on cooperation, community support, volunteer energy, and public funds.

Another way the GFB attempts to reclaim the commons is by resisting geographically expansive global commodity chains and shortening the links between production and consumption, or what FoodShare calls "field to table." FoodShare's buying practices support bioregional foodsheds against practices of corporate agribusiness and long-distance global trade. While not all of the produce is locally grown, even in the regular (nonorganic) GFB there is a much higher percentage of local produce than one would find in a typical supermarket. In some summer months, the regular box is primarily locally produced and often contains a substantial percentage of organic produce. Through these buying practices, the GFB supports sustainable agriculture in southern Ontario; farmers are not seen as "labor inputs" in the supply process but as ecological stewards, and they are encouraged through information exchanges, price supports, and collective meetings to discuss needs and opportunities.

Of course, there is no absolute sense in which FoodShare can reclaim food commons from extensive commodification. Like the Fair Trade projects Shreck discusses in the next chapter, dependence on the market involves a series of contradictions. The GFB tries to limit the market and reclaim food as a basic right, but the market in turn works to limit the counterhegemonic aspirations of the GFB. This happens through entrenched poverty and social inequality, which seriously limit access to the GFB program. While the GFB aspires to support local organic farmers, this goal rubs up against its other goal of providing affordable food for low-income people (see Allen 1999). The GFB is still too expensive for most of the poorest and hungriest people using Toronto's food banks—an estimated 160,000 each month—in addition to an additional 160,000 who are food insecure but do not use food banks. As with Fair Trade initiatives, efforts at income redistribution are only partially successful, as they come up against the severe income inequality that exists within the globalized capitalist system.

While there is pressure to make the GFB inexpensive enough to be accessible to low-income people, there are limits to how cheaply FoodShare can price the box without compromising its commitments to farmers' livelihoods. This is a built-in tension that is hard to avoid or remedy in a single food project, and it worsens depending on the precariousness of organizational funding—usually a considerable strain for nonprofit organizations.[9] This tension has increased with the global expansion of commodity chains (Kneen 1993), and the further vertical and horizontal concentration of the retail industry (Guptill and Wilkins 2002). Former warehouse manager Mary-Lou Morgan observed that it had become increasingly difficult to price the GFB competitively; discount grocery chains don't deal with the Ontario Food Terminal but have direct supply connections that eliminate wholesalers and

9. One result of this desire to make FoodShare more financially "sustainable" is the never-ending search for new sources of funding. The "Field-to-Table" festival, for example, was in part sponsored by RBC Financial Group, one of Canada's largest financial institutions.

allow for deeply discounted prices. Morgan wondered how long the GFB could last in this fiercely competitive environment.

This brings us to the second criterion of counterhegemonic food politics: creating postconsumer needs and desires. The GFB brings local fruits and vegetables to urban eaters, but this involves more than servicing customers. Most important, the "consumer" does not exercise complete sovereignty in this relationship. Individual consumer agency is limited for the sake of building collective empowerment through a collective project committed to sustainable, democratic, equitable outcomes. Individual participants in the GFB program do not control what goes into the box; they voluntarily limit their "right" to choice by signing up for a standardized box—something very difficult, if not impossible, for a for-profit delivery service. While attempts are made to keep people happy (e.g., one substitute may be permitted per order), customers have to be satisfied with what is local, seasonal, available, and affordable. This standardization is based on a sense that consumers have individual rights but that they are also citizens with collective responsibilities—to farmers, to the land, and to eating what is in season. Through a weekly newsletter and the interaction with FoodShare events and staff, the production-consumption chasm lessens as urbanites learn about the farmers who grow their food. This minimizes the anthropocentric tendency to separate production from consumption, presents food producers as ecological stewards, and de-fetishizes food as a commodity to be sought out as cheaply as possible by individual consumers.

A noncommodified relationship and sense of pleasure can also be observed in the volunteers and youth workers who pack the GFB. The relationship between the workers and the food is not a strict labor transaction but at least in part is a labor of love and self-respect.[10] Volunteers are mainly low-income workers who get a free box at the end of the day. Members of the focus-on-youth training program contribute substantial labor to the program, carrying out the bulk of the ordering and organizing, and often intervening to make sure that the produce is not bruised or spoiled. There is a tangible sense of pleasure and pride that connects the workers to the food. In the midst of the exceptional busyness and outright chaos of packing day, volunteers gather the workers around the assembly line and talk about the food, where it came from, and the farmers who grew it. Farmers are occasionally

---

10. Despite the upbeat and respectful atmosphere found in the FoodShare warehouse, it is important to note that the "choice" to provide free labor is limited when one is an unemployed person forced to volunteer by bureaucratic regulations, or a low-income person left hungry by cuts to social services.

present to share their own stories, and workers often contribute their thoughts on the foods or preparations techniques used in their "home" countries. At such moments food is not a simple commodity but an intimate good linked to a sense of self, identity, and pleasure shared with others.

Of course, there are limits on creating postconsumer needs and pleasures through the GFB program. People at all income levels still want value for their money. Weaning passive consumers off cheap industrial food is not easy, particularly when it costs less than the organic, local, or fairly traded alternative. According to market research conducted for FoodShare, one reason that more people do not sign up for the GFB is that they want to retain control over individual choice in the grocery store. Enrollment is thereby limited by the fact that people feel it is their right as consumers to have choice, and it is possible that this choice is vital to individuals who experience minimal power to control the outcomes in other domains of their lives. The GFB customers who accept the standardization of products do so because the box is relatively affordable and convenient; motivations to reduce social or ecological exploitation are not necessarily paramount.[11]

It is also not clear to what extent GFB subscribers take on an identity beyond "consumer." They often refer to themselves as consumers, they call the office and complain like consumers, and even the FoodShare staff uses the discourse of consumerism. Although more research is needed on this question, it does not always appear that individual participants see themselves as members of a collective food security social movement or even as members of FoodShare as an organization. This issue is apparent in the FoodShare annual general meeting (AGM) held each spring. While several hundred people attend each year, partly because of the popular plant sale, which offers interesting varieties of heritage bedding plants, thousands of volunteers and GFB participants do not show up. To achieve a quorum in the voting process, FoodShare needs a certain number of members to vote at the AGM. Even though membership is free, people often become suspicious or anxious when asked to join and to attend the AGM, and seem to be worried about attendant responsibilities and commitments. This suggests that the GFB exists in a cultural space still governed by the laws and norms of consumer sovereignty—a realm with more individual consumer rights than collective empowerment. The recent emphasis within FoodShare on the "Field-to-Table" educational campaigns attempts to build a more popular sense of food politics and organizational identity for FoodShare among Torontonians.

11. One internal FoodShare study found that only 5 percent of respondents cited interest in helping farmers as a motive, and this 5 percent fell into the wealthy subset of GFB clients.

In sum, while the GFB does not reach all the hungry people in the city of Toronto, it does provide an empowered model of food provisioning that combines small-scale markets with cooperative, nonprofit energies to facilitate community empowerment. While it encourages local markets, the GFB does not promote commodification and profit maximization at the expense of other social goals and needs. According to the criteria I put forward, the GFB does attempt to move beyond a hegemonic system of industrial food production and consumption (see Table 4.2). FoodShare's GFB restricts the domain of the market in food production and attempts to shift discourse and practices away from food as a commodity toward food as a basic human need and something that should be sustainably grown and available to people irrespective of income. The GFB also strives to shorten the distance between production and consumption of food through its "Field-to-Table" approach, which builds connections with area growers. Finally, the GFB challenges individual consumer sovereignty, values ecological stewardship (not just individual consumer choice), invites its participants to rethink their food preferences (e.g., challenging the desire for fresh strawberries in January),

Table 4.2

Counterhegemonic Criteria	FoodShare's GFB Response	Limitations
1. Reclaiming the Commons	Assert food as a human need, not simply a commodity.	Social inequality limits accessibility, particularly of organic food.
	The end goal is not profit maximization but the use of a nonprofit community-based organization to increase food access, build community empowerment, and shorten the distance from field to table.	Global commodity chains lower the cost of conventional long-distance agriculture, making it difficult for the GFB to compete.
2. Creating Postconsumer Needs and Desires	The GFB restricts food choices to the local, seasonal, and affordable and challenges the notion of consumer sovereignty at the expense of ecological sustainability.	Large numbers of consumers value maximum choice and minimum price in food purchases—especially if the ecological and social price tag is not transparent.
	The pleasure of eating local, seasonal, and socially just food is developed for workers and consumers.	

and creates a collective project striving for more democratic, sustainable out-comes in the food system. The adjudication criteria suggest that the GFB pro-gram provides aspects of a counterhegemonic alternative to corporate food chains and at the same time identifies the limits to empowerment built into the hegemonic culture of consumerism and the exceptional power of mar-ket forces to transport cheap industrial food along global commodity chains.

The adjudicating framework can also provide a way of comparing the GFB to other programs, such as food banks, for-profit delivery services, and the rapidly expanding corporate interest in organics and "natural" foods. One particularly interesting comparison involves the expanding phenome-non of "organic" superstores. The U.S. chain Whole Foods Market is now set to become one of the five hundred largest businesses in the country. Its goal is to double the number of stores to three hundred and bring sales up to $10 billion by 2010 (Gertner 2004). Even a cursory comparison of the GFB with a similar basket of produce purchased at a Whole Foods Market using these adjudication criteria suggests important differences. Unlike the GFB, Whole Foods does not attempt to decommodify food production but explicitly aspires to maximize profitability and expand market share. While some local products are available in Whole Foods stores, variety is prioritized over seasonality (a strategy that is heavily reliant on the low-cost availability of California organics subsidized by cheap fossil fuels), and the store is orga-nized around a system of centralized distribution through chain stores. It is driven by a model of consumer sovereignty that does not restrict choice for reasons of seasonality or locality; shoppers are encouraged to *expand* their individual food wants, not restrict them for the sake of collective ecological and social goals. The Whole Foods strategy privileges a highly individual-ized commodity response to the ill health of common food resources, and does not seriously challenge a corporate production system that prioritizes profits over ecological integrity or social justice.

This comparison is not intended to demonize Whole Foods or to mini-mize some of its unique innovations,[12] but to indicate the unique qualities of the GFB and the structural limitations of for-profit solutions based exclu-sively on individual commodity consumption. While FoodShare is limited by the market economy in which it must operate, its role as a nonprofit organization, a community-building project, and a publicly supported in-stitution allows it take on a more counterhegemonic role—challenging the

12. One Whole Foods policy bars CEOs from making more than fourteen times the average company salary; at the same time, the corporation and its CEO, John Mackey, are famous for their antiunion stance.

hegemony of consumerism, prioritizing ecological integrity over unlimited consumer choice, providing a sense of food's pleasure through cooperative community-building practices, and asserting that healthy food is a basic life-good, not a simple commodity.

Conclusion

While buying organic food does not automatically make one a revolution-ary, blanket pronouncements on bourgeois piggery are no substitute for sociological research on food politics. Hegemony is constantly being re-negotiated, and counterhegemonic potential can be identified when actors (1) reclaim and invent civil commons traditions to broaden access to basic needs like food, restricting the commodity system that separates production and consumption; and (2) promote new "postconsumer" needs and desires for food that is not just delicious but is rooted in empowered relationships of social justice and ecological integrity.

Developing concrete criteria for the study of the counterhegemonic potential of food-specific politics is important for at least two reasons. First, it is important for theory building. Leftist social theory and political ecol-ogy have excelled at critiques of postindustrial capitalism, while neglecting the task of identifying empowerment and building alternatives. Soper writes that "in matters concerning poverty, injustice, and ecological devastation, it is very much easier to expose the sins and sinners than it is to point to the means and the agents of correction" (1999, 118). While FoodShare's GFB is not a panacea to the problems of industrial agriculture, programs like the GFB pro-vide important lessons for identifying "agents of correction" and pragmatic examples of how empowerment exists in real-life political-economic con-texts. However impressive these microprojects of empowerment, the agency of FoodShare is limited by the greater power of the agrifood industrial com-plex and the contradictions of its market dependency. Future research must investigate how to extrapolate successful small-scale models like the GFB at broader scales, without creating alienating bureaucratic hierarchies and dis-tanced food sources, all the while appealing to urban eaters steeped in the paradigm of consumer sovereignty (see Johnston and Baker 2005).

This suggests a second reason for studying the counterhegemonic poten-tial of food politics: the development of social policy alternatives. If we sim-ply dismiss burgeoning interest in healthy, organic food as elitist, we lose the ability to discriminate and identify progressive policy possibilities—like a GFB program supported federally, or publicly supported farmers' market programs, food appreciation classes in public schools, or a healthcare card that could be used to acquire food staples like fruits and vegetables. We also lose the ability to criticize hegemonic corporate responses to collective anxieties about unhealthy food and landscapes. Given that hegemony is a never-ending battle for cultural leadership, there will always be innovative new strategies emerging from corporate boardrooms to commodify countercultural impulses (Frank 1997), and food is no exception (Belasco 1993). While academic theorizing suffers a bad reputation for being insular and self-serving, good social theory always combines the theoretical with the empirical. As food researchers, this focus keeps our eye on the ball, enabling us to identify unsustainable food systems and encourage culinary impulses based on social justice and ecological stewardship.

REFERENCES

Adkin, Laurie. 1992. "Counter-Hegemony and Environmental Politics in Canada." In *Organizing Dissent: Contemporary Social Movements in Theory and Practice,* ed. William Carroll, 135–56. Toronto: Garamond Press.

Allen, Patricia. 1999. "Reweaving the Food Security Safety Net: Mediating Entitlement and Entrepreneurship." *Agriculture and Human Values* 16 (2): 117–29.

Belasco, Warren J. 1993. *Appetite for Change: How the Counterculture Took on the Food Industry.* Ithaca: Cornell University Press.

Bourdieu, Pierre. 1984. *Distinction: A Social Critique on the Judgment of Taste.* London: Routledge.

Bove, José, and François Dufour. 2001. *The World Is Not for Sale: Farmers Against Junk Food.* Interviewed by Gilles Luneau. Translated by Anna De Casparis. London: Verso.

Buck, Daniel, Christina Getz, and Julie Guthman. 1997. "From Farm to Table: The Organic Vegetable Commodity Chain of Northern California." *Sociologia Ruralis* 37 (1): 3–20.

Campell, David. 2001. "Conviction Seeking Efficacy: Sustainable Agriculture and the Politics of Co-optation." *Agriculture and Human Values* 18 (4): 353–63.

Carroll, William. 1992. "Introduction: Social Movements and Counter-Hegemony in a Canadian Context." In *Organizing Dissent: Contemporary Social Movements in Theory and Practice,* ed. William Carroll, 1–21. Toronto: Garamond Press.

Conway, Janet. 2004. *Identity, Place, Knowledge: Social Movements Contesting Globalization.* Halifax, Canada: Fernwood.

Critser, Greg. 2003. *Fat Land: How Americans Became the Fattest People in the World.* Boston: Houghton Mifflin.

Crotty, Pat. 1999. "Food and Class." In *Sociology of Food and Nutrition: The Social Appetite,* ed. John Germov and Lauren Williams, 135–48. Oxford: Oxford University Press.

Duncan, Colin. 1996. *The Centrality of Agriculture: Between Humankind and the Rest of Nature.* Montreal: McGill-Queen's University Press.

DuPuis, E. Melanie. 2000. "Not in My Body: rBGH and the Rise of Organic Milk." *Agriculture and Human Values* 17 (3): 285–95.

Drake, Mary Anne. 1992. "The Nutritional Status and Dietary Inadequacy of Single Homeless Women and Their Children in Shelters." *Public Health Reports* 107 (3): 312–19.

Ehrenreich, Barbara. 1985. *The Worst Years of Our Lives.* New York: Pantheon Books.

———. 1994. "Berkeley." *Z Magazine,* September, 5.

Foucault, Michel. 1977. *Discipline and Punish: The Birth of the Prison.* Trans. C. Gordon. New York: Harvester Wheatshaff.

———. 1990. *The History of Sexuality.* Vol. 1. Trans. R. Hurley. New York: Penguin Books.

Frank, Thomas. 1997. *The Conquest of Cool.* Chicago: University of Chicago Press.

Freire, Paulo. 1970/2000. *Pedagogy of the Oppressed.* New York: Continuum.

Friedland, William H. 1994. "The New Globalization: The Case of Fresh Produce." In *From Columbus to ConAgra: The Globalization of Agriculture and Food,* ed. Alessandro Bonnano, Lawrence Busch, William H. Friedland, Lourdes Gouveia, and Enzo Mingione, 210–31. Lawrence: University Press of Kansas.

Gertner, Jon. 2004. "The Virtue in $6 Heirloom Tomatoes." *New York Times Magazine,* 6 June. www.nytimes.com/.

Giddens, Anthony. 1991. *Modernity and Self-Identity: Self and Society in the Late Modern Age.* Cambridge, UK: Polity Press.

Goldman, Michael. 1998. "Introduction: The Political Resurgence of the Commons." In *Privatizing Nature: Political Struggles for the Global Commons,* ed. Michael Goldman, 1–19. New Brunswick: Rutgers University Press.

Gramsci, Antonio. 1971. *Selections from the Prison Notebooks.* Ed. and trans. Quintin Hoare and Geoffrey Nowell Smith. New York: International Publishers.

Guptill, Amy, and Jennifer Wilkins. 2002. "Buying into the Food System: Trends in Food Retailing in the U.S. and Implications for Local Foods." *Agriculture and Human Values* 19 (1): 39–51.

Gusfield, Joseph R. 1992. "Nature's Body and the Metaphor of Food." In *Cultivating Differences: Symbolic Boundaries and the Making of Inequality,* ed. Michele Lamont and Marcel Fournier, 75–103. Chicago: University of Chicago Press.

Guthman, Julie. 1998. "Regulating Meaning, Appropriating Nature: The Codification of California Organic Agriculture." *Antipode* 30 (2): 135–54.

———. 2002. "Commodified Meanings, Meaningful Commodities: Re-thinking Production-Consumption Links Through the Organic System of Provision." *Sociologia Ruralis* 42 (4): 295–311.

Hall, Stewart. 2001. "Cultural Studies." In *The New Social Theory Reader,* ed. Steven Seidman and Jeffrey Alexander, 88–100. New York: Routledge.

Harvey, David. 1989. *The Postmodern Condition.* Malden, Mass.: Blackwell.

Hawken, Paul, Amory Lovins, and L. Hunter Lovins. 2000. *Natural Capitalism.* Boston: Little, Brown.

Hinrichs, C. Clare. 2000. "Embeddedness and Local Food Systems: Notes on Two Types of Direct Agricultural Markets." *Journal of Rural Studies* 16 (3): 295–303.

Homer-Dixon, Thomas. 2001. *The Ingenuity Gap: Can We Solve the Problems of the Future?* Toronto: Vintage Canada.

Jackson, Peter. 1993. "A Cultural Politics of Consumption." In *Mapping the Futures: Local Cultures, Global Change,* ed. John Bird, Barry Curtis, Tim Putnam, G. Robertson, and Lisa Tickner, 207–28. New York: Routledge.

Johnston, Josée. 2002. "Consuming Social Justice: Fair Trade Shopping and Alternative Development." In *Protest and Globalisation: Prospects for Transnational Solidarity*, ed. James Goodman, 38–56. Melbourne, Australia: Pluto Press.

———. 2003. "Who Cares About the Commons?" *Capitalism, Nature, Socialism* 14 (3): 1–41.

Johnston, Josée, and Lauren Baker. 2005. "Eating Outside the Box: FoodShare's Good Food Box and the Challenge of Scale." *Agriculture and Human Values* 22 (3): 1–13.

Kauffman, L. A. 1991. "New Age Meets New Right: Tofu Politics in Berkeley." *The Nation*, 16 September, 294–96.

Keat, Russell. 1999. "Market Boundaries and the Commodification of Culture." In *Culture and Economy After the Cultural Turn*, ed. Larry Ray and Andrew Sayer, 92–111. London: Sage Publications.

Kempson, Kathryn, Debra Palmer Keenan, Puneeta Sonya Sadani, Sylvia Ridlen, and Nancy Scotto Rosato. 2002. "Food Management Practices Used by People with Limited Resources to Maintain Food Sufficiency as Reported by Nutrition Educators." *Journal of the American Dietetic Association* 102 (12): 1795–99.

Kloppenburg, Jack R., Jr., Sharon Lezberg, Kathy DeMaster, G. W. Stevenson, and John Hendrickson. 2000. "Tasting Food, Tasting Sustainability: Defining the Attributes of an Alternative Food System with Competent, Ordinary People." *Human Organization* 59 (2): 177–86.

Kneen, Brewster. 1993. *From Land to Mouth: Understanding the Food System*. Toronto: NC Press.

———. 1999. "Restructuring Food for Corporate Profits: The Corporate Genetics of Cargill and Monsanto." *Agriculture and Human Values* 16 (2): 161–67.

Kovel, Joel. 2002. *The Enemy of Nature: The End of Capitalism or the End of the World?* Nova Scotia: Fernwood Publishing.

Levenstein, Harvey. 1993. *Paradox of Plenty: A Social History of Eating in Modern America*. New York: Oxford University Press.

Mallick, Heather. 2002. "When I Go, I Want My Slippers On." *Globe and Mail* (Toronto), 30 November, F2.

Mano, Jack. 2002. "Commoditization: Consumption Efficiency and an Economy of Care and Connection." In *Confronting Consumption*, ed. Thomas Princen, Michael Maniates, and Ken Conca, 67–100. Cambridge: MIT Press.

McIntyre, Linda. 2003. "Food Security: More Than a Determinant of Health." *Policy Options* (March): 46–51.

McMurtry, John. 1998. *Unequal Freedoms: The Global Market as an Ethical System*. Toronto: Garamond Press.

———. 1999. *The Cancer Stage of Capitalism*. Sterling, Va.: Pluto Press.

———. 2001. "The Life-Ground, the Civil Commons, and the Corporate Male Gang." *Canadian Journal of Development Studies* 22 (4): 819–44.

Mennell, Stephen. 1986. *All Manners of Food: Eating and Taste in England and France from the Middle Ages to the Present*. New York: Basil Blackwell.

O'Connor, James. 1996. "The Second Contradiction of Capitalism." In *The Greening of Marxism*, ed. Ted Benton, 197–221. New York: Guilford Press.

Riches, Graham. 2002. "Food Banks and Food Security: Welfare Reform, Human Rights, and Social Policy; Lessons from Canada?" *Social Policy and Administration* 36 (6): 648–63.

Roseberry, William. 1994. "The Language of Contention." In *Everyday Forms of State Formation: Revolution and the Negotiation of Rule in Modern Mexico,* ed. Gilbert M. Joseph and Daniel Nugent, 355–65. Durham: Duke University Press.

Sachs, Wolfgang. 1998. *Greening the North: A Post-Industrial Blueprint for Ecology and Equity.* London: Zed Books.

Schaap, Andrew. 2000. "Power and Responsibility: Shall We Spare the King's Head?" *Politics* 20 (3): 129–35.

Seccombe, Wally. 2002. "NGOs and Social Democracy." Paper presented at the Future of Social Democracy in Canada Conference, sponsored by the McGill Institute for the Study of Canada, Montreal, 25–26 May. Available at www.arts.mcgill/ca/programs/misc/socdem/secco.htm.

Shreck, Aimee. 2002. "Just Bananas? Fair Trade Banana Production in the Dominican Republic." *International Journal of Sociology of Agriculture and Food* 10 (2): 11–21.

Sklair, Leslie. 2001. *The Transnational Capitalist Class.* Oxford: Blackwell.

Slater, David. 1997. *Consumer Culture and Modernity.* Oxford: Polity Press.

Smith, Philip. 2001. *Cultural Theory: An Introduction.* Malden, Mass.: Blackwell.

Smith, Toby. 1998. *The Myth of Green Marketing: Tending Our Goats at the Edge of Apocalypse.* Toronto: University of Toronto Press.

Sobal, Jeffery, and A. J. Stunkard. 1989. "Socioeconomic Status and Obesity: A Review of the Literature." *Psychological Bulletin* 105 (2): 260–75.

Soper, Kate. 1990. *Troubled Pleasures: Writings on Politics, Gender, and Hedonism.* New York: Verso.

———. 1999. "Other Pleasures: The Attractions of Post-Consumerism." In *Necessary and Unnecessary Utopias: Socialist Register 2000,* ed. Leo Panitch and Colin Leys, 114–32. New York: Monthly Review Press.

Tarasuk, Valerie, and Barb Davis. 1996. "Responses to Food Insecurity in the Changing Canadian Welfare State." *Journal of Nutrition Education* 28 (2): 71–75.

Van der Pijl, Kees. 1998. *Transnational Classes and International Relations.* London: Routledge.

Veblen, Thorstein. 1899/1994. *The Theory of the Leisure Class.* New York: Penguin Books.

Walther, Gian-Reto, Eric Post, Peter Convey, Annette Menzels, Camille Parmesan, Trevor Beebee, Jean-Marc Fromentin, Ove Hoegh-Guldberg, and Franz Bairlain. 2002. "Ecological Responses to Recent Climate Change." *Nature* 416: 389–95.

Wilson, E. O. 1992/1999. *Diversity of Life.* New York: W. W. Norton.

Winson, Anthony. 1993. *The Intimate Commodity.* Toronto: Garamond Press.

Wood, Elaine Meiksins. 1999. *The Origin of Capitalism.* New York: Monthly Review Press.

INSTRUCTOR'S RESOURCES

Key Concepts and Terms:
1. Hegemony
2. Counterhegemony
3. Food Politics
4. Community Food Security

Discussion Questions:
1. What are three ways that consumers can seek positive change by altering their shopping habits? Can these be considered counterhegemonic?

2. Is targeting individual consumption habits an effective strategy for social change? Does a focus on individual consumption let governments and corporations off the hook for enforcing social justice and ecological preservation?

3. Should small-scale community food projects, like FoodShare's good food box, be expanded to reach a greater number of people? If so, how could this happen?

Agriculture, Food, and Environment Video:

1. *The Corporation.* Mark Achbar and Jennifer Abbott, 2004 (145 minutes). See www.thecorporation.tv/.

Agriculture, Food, and Environment on the Internet:

1. FoodShare: www.FoodShare.net/.

2. Community Food Security: www.foodsecurity.org/.

3. Toronto Food Policy Council: www.city.toronto.on.ca/health/tfpc_index.htm.

Additional Readings:

1. For an overview of contemporary environmental trends, see the United Nations Environmental Program's *Global Environmental Outlook,* vols. 1–3, which document the critical crossroads of humanity's next thirty years on the planet: www.unep.org/geo/.

2. For a general overview of the Gramscian concept of hegemony, see the short volume by Robert Bocock, *Hegemony* (New York: Routledge, 1986).

3. For a fascinating and original account of the incorporation of countercultural impulses into corporate marketing tools, see Thomas Frank's *The Conquest of Cool* (Chicago: University of Chicago Press, 1997).

4. To learn more about the history of food and the counterculture in the United States, see Warren J. Belasco, *Appetite for Change: How the Counterculture Took on the Food Industry* (Ithaca: Cornell University Press, 1993).

5

RESISTANCE, REDISTRIBUTION, AND POWER IN THE FAIR TRADE BANANA INITIATIVE

Aimee Shreck

The Fair Trade movement critiques the conventional agrifood system by connecting producers in the global South with consumers in the global North through alternative trade channels that are more equitable than those typical of conventional trade networks (Murray and Raynolds 2000; Raynolds 2002; Renard 2003). Worldwide, sales of Fair Trade–labeled products (including commodities like coffee, tea, sugar, honey, and bananas) are increasing. For example, they rose by more than 20 percent between 2000 and 2001. The Fairtrade Labelling Organizations International (FLO), the umbrella organization that sets Fair Trade standards and provides oversight to the Fair Trade network, currently works with 315 certified producer groups who represent some nine hundred thousand families of small farmers and workers in the South (FLO 2003). In recognition of its work, FLO was recently awarded the prestigious International Prize for Development from the King Baudouin Foundation in Belgium. Accomplishments such as these draw attention to the work of Fair Trade organizations and invite inquiry into whether such work is actually a "good deal" for the farmers it purports to assist.

Potential Fair Trade advocates, as well as those with less enthusiasm, want to know if and why they should buy Fair Trade–labeled products and wonder how oppositional the Fair Trade movement is. There are no simple, straightforward answers to these questions. As Murray, Raynolds, and Taylor note, summary conclusions about Fair Trade initiatives may fail to capture the subtleties of the promises and obstacles embedded in the movement: "Both celebratory and less sanguine accounts of Fair Trade abound in popular and scholarly literature. Yet the questions underpinning the assessments

An earlier version of this chapter appeared in *Agriculture and Human Values* 22, no. 1 (2005): 17–29.

of Fair Trade are often more complex and the answers more ambiguous, than many of these accounts recognize" (2003, 1).

This line of inquiry is not restricted to Fair Trade initiatives. Rather, as the other chapters in this volume demonstrate, questions are increasingly being asked about the broader potential of other alternatives and countermovements (e.g., community food security coalitions and community-supported agriculture) that have emerged to confront the globalizing tendencies of the conventional agrifood system.[1] For example, in her analysis of local food system initiatives in Iowa, Hinrichs asks, "What is the transformative potential of the current efforts to promote production and consumption of foods earmarked by locality or region?" (2003, 33). In doing so, she addresses how localization efforts that grow out of opposition to homogenization, industrialization, and concentration in the global food system can arguably be characterized as liberatory and/or reactionary. Allen et al. ask related questions in their examination of thirty-seven alternative agrifood initiatives in California, each of which works in its own way to challenge the existing food system and build alternatives. Their research assesses whether and how the initiatives are oppositional or alternative. In their words, they are concerned with understanding the degree to which these alternatives "seek to create a new structural configuration . . . and to what degree . . . their efforts [are] limited to incremental erosion at the edges of the political-economic structures" (Allen et al. 2003, 61). Johnston's research, which questions and identifies the counterhegemonic characteristics of a community food security initiative in Canada, is similarly framed around a concern with the wider implications of local food system initiatives that skeptics might be tempted to write off as examples of "bourgeois piggery" (see previous chapter).

This case study of Fair Trade also addresses such questions of agency, resistance, and transformative potential by drawing attention to the multiple levels at which Fair Trade operates. I consider the ways in which the Fair Trade movement encourages actors to engage in different forms of social action to clarify the movement's counterhegemonic potentials, where they are being realized, and where they are not. I suggest that the movement is most successful in enabling consumers and producers to commit acts of resistance and in facilitating the redistribution of resources from the North to the South. Up to now, though, Fair Trade alternatives appear to hold only a theoretical potential to provoke more transformative change in the agrifood

1. McMichael (2000b, 22) describes a "plethora of alternatives—including community supported and sustainable agriculture, community food security coalitions, organic food, principles of biodiversity, vegetarianism, fair trade movements, eco-feminism."

system. The chapter concludes with a suggestion that a reconceptualization of the model upon which Fair Trade initiatives are built and the way in which they are implemented could allow the movement to realize more of its oppositional promise.

Fair Trade Bananas: An "Experience in Progress"

The alternative trade organizations (ATOs) promoting and practicing Fair Trade today generally trace their roots to the 1960s, when church-based and development organizations, primarily in Europe, set up shops to sell handicrafts and similar items purchased directly from Third World producers.[2] These "Fair Trade" products were sold to consumers who supported this form of solidarity with producers and workers from politically marginalized countries. During the next two decades, ATOs organized themselves and formed international networks. Alternative trading started more slowly in the United States, but the movement has recently begun to take off more forcefully (Conroy 2001; Zonneveld 2003). Internationally, the practice of Fair Trade was formalized with the introduction of Fair Trade labels in 1988 and has become more standardized with the establishment of FLO in 1997 to set and manage Fair Trade standards and coordinate the activities of its seventeen member organizations.[3] Today, a sophisticated system of standards and certification and a common Fair Trade label are in place to support the Fair Trade initiatives.

These changes reflect how the movement itself has developed over time, as it has been faced with difficult choices about the Fair Trade strategy and the movement's objectives. As Renard describes, prior to the introduction of labels, some advocates were concerned that alternative trade "questioned the mechanisms of the dominant market system and proposed a fairer relationship between producers and consumers" but "was far from resolving the problems of selling Third World products" (2003, 89–90). To really increase sales opportunities of Fair Trade products, ATOs saw that products would have to be more widely available. But this move implied working within large distribution channels and required a change in the organization's message. That

2. Renard's (2003, 87) apt description of Fair Trade as an "experience in progress" is particularly fitting for the Fair Trade banana initiative, which is dynamic and has been evolving in response to its stakeholders' needs from the start.

3. Currently, members include fourteen national initiatives from Europe, plus initiatives in Canada, the United States, and Japan.

is, "to broaden the spectrum of the public interested in buying these products, it was necessary to appeal more to humanitarian sentiments than to political convictions" (90). It is important to recognize that the decision to make this shift and to rely on conventional circuits of distribution came only after months of difficult discussion and strong resistance.

This tension of working simultaneously "in and against" the market (Brown 1993), formalized with this key step, continues to be highlighted as a sensitive issue in discussions about Fair Trade, both internally (e.g., Gereffi 2000; Zonneveld 2003) and externally (e.g., Raynolds 2002; Renard 2003). Since the strategy is based on working within the same global capitalist market that it hopes to alter, it raises questions about whether Fair Trade initiatives actually hold any counterhegemonic potential to transform the conventional agrifood system into something qualitatively different and presumably better.

This history and the broad appeal of a concept like "fair trade" permits the term to be used in various ways. My understanding of the term is consistent with the work of Murray and Raynolds (2000) and with the way in which Fair Trade organizations conceptualize themselves and their activities. I view Fair Trade as a contemporary social movement that contests the conventional agrifood system and rejects the exploitive social and environmental relations of production that characterize it. In comparison with other oppositional agrifood movements (e.g., organic agriculture, GMO labeling campaigns, community food security), the Fair Trade movement is unique in its attempt to go beyond improving production conditions to actually alter conventional trade relations (Raynolds 2000, 2002). Like these other movements, the Fair Trade movement is made up of individuals and groups engaged in organized collective action that challenges the dominant corporate food regime (McMichael 2000b).

Yet, despite the increasing size and momentum of movements like Fair Trade, the extent to which they are actually disrupting the global political economic system into which they fit is not always clear. As Evans has noted, the increasing resistance to globalization that we are witnessing is not by definition oppositional. In his words, "the surprising resilience and adaptive ability of ordinary people whose lives have become transnational does not necessarily challenge the dominant global rules, the way these rules are made, or the economic ideology that legitimates them" (2000, 230).

With this observation in mind, my analysis builds on insights and examples from field research conducted on the banana initiative, with specific attention to the bananas produced in the Dominican Republic and exported to Europe.[4] The majority of the fieldwork was conducted in ten communities in the Azua Valley of the Dominican Republic with Fair Trade–certified banana farmers during a period of eight months in 2000 (Shreck 2002a).[5] This region is home to approximately eight hundred small-scale banana farmers, roughly two-thirds of whom belonged to certified Fair Trade associations at the time. These banana producers were among the first to be certified by FLO (in 1997), and represent the largest group of small-scale producers exporting Fair Trade bananas. Significantly, their experiences reflect the relative youth and inexperience of the Fair Trade banana initiative.

Primary data were collected during semistructured interviews with a random sample (n = 115) of small-scale banana producers representing three of the main producer groups in the region. Two of these groups were/are registered with FLO as Fair Trade producer associations. All of the bananas grown by these producers were also certified and sold as organic. Additional data from qualitative interviews with exporters, Fair Trade organization representatives, and other local experts and key informants, as well as participant and nonparticipant observations (e.g., at meetings, in communities, during packing days) further inform this analysis.

Since most of the discussions about Fair Trade draw primarily on the Fair Trade coffee initiative, it is important to appreciate that although the basic principles of Fair Trade are the same for each commodity, there are certain characteristics that are unique to each product. While coffee has traditionally been at the center of the Fair Trade movement's activities and is the most widely consumed Fair Trade commodity, bananas are increasingly important for Fair Trade organizations. Since Fair Trade bananas were introduced in 1996, they have captured unprecedented market shares, reaching more than 20 percent in Switzerland (FLO 2002, 2003). Perhaps the most important distinction of bananas is that they are the first fresh fruit commodity that Fair Trade organizations attempted to certify. Applying the Fair Trade model to a highly perishable, fresh fruit product places Fair Trade squarely in one of the most competitive and globalized sectors of the agrifood system (Friedland 1994) and presents new logistical challenges for the organizations to manage. Because of this position, Fair Trade banana traders must rely more heavily on conventional methods and intermediaries for shipping and storage than do traders of nonperishable products.

---

4. At that time, Fair Trade bananas were sold only in Europe.

5. It is important to note that my analysis reflects this timeframe. Since that time there have been developments and changes in FLO—in the banana initiative particularly—and within the producer organizations.

The remainder of this chapter proposes a preliminary framework for assessing the oppositional promise of Fair Trade initiatives that takes into account the complexity of a movement operating at multiple levels, across long distances, and with a diverse network of actors. Though the findings should not be generalized to all Fair Trade commodities, this case can provide insights into how to think about and evaluate the potentials and limits of other initiatives.

Fair Trade as Counterhegemonic Social Action

Contemporary transformations of the agrifood system reflect the same tendencies and processes that are characteristic of the dynamics in the global economy. Accordingly, the globalization of agriculture and food is rooted in capitalism and hence motivated by profit and efficiency, driven by transnational corporate actors, and upheld by the dominant ideology of neoliberalism (Held et al. 1999; McMichael 2000b). Efforts to liberalize agriculture include reducing farm subsidies and other protections for agricultural sectors, and are being institutionalized through international free trade agreements and the World Trade Organization (WTO). More specifically, the hegemonic forces that uphold the dominant agrifood system are the key elements of what McMichael (2000a) describes as a global corporate food regime that is characterized by tendencies toward increasing centralization, concentration, corporate control, and privatization (Grey 2000; Heffernan 2000; Magdoff, Foster, and Buttel 2000). Under this regime, world agriculture is being restructured in a way that sharpens divisions between the North and the South.

Yet diverse and widespread opposition to this globalized, capitalist agrifood system signals that the system is unsatisfactory in many ways and is linked to a host of social, environmental, and economic problems. Indeed, the industrial model of agriculture privileged in this system is kept profitable by exploitive and destructive labor and environmental practices. Moreover, the policies through which it is institutionalized and legitimated are designed to deliver political and material benefits to corporate agribusiness, to the disadvantage of small farmers and workers around the world.

The responses to the negative effects of this system are global as well. Not surprisingly, academic attention to the outpouring of resistance and alternatives is lively, multidisciplinary, and continuously growing (e.g., Marsden 2000; Raynolds 2000; Allen et al. 2003), as researchers consider how to understand these countermovements. To clarify the possibilities for Fair Trade initiatives to foster positive change and reshape the agrifood system, I question the extent to which the Fair Trade movement could be considered counterhegemonic, discussing where it holds (and realizes) counterhegemonic potential.

My understanding of counterhegemony draws on Gramsci's (1971) insights about the complexity of political struggle and his argument that a successful strategy for transformative change in society must include multiple forms of action. In particular, his emphasis on the strength of a "war of position" strategy, in which multiple forms of action are taken at many sites of struggle to confront a hegemonic system indirectly, is helpful for unpacking the different ways in which Fair Trade actors work together to make their initiatives successful.

More recently, Chin and Mittelman (2000) discuss "resistance as counterhegemony" in terms of antiglobalization movements. They remind us that not only are forms of resistance multiple but that the agents of resistance are diverse, that sites of resistance are found at different levels, and that forms of struggle are many. Applying Gramsci's understanding of the possibilities emerging out of a "war of position" to the present-day struggle against corporate globalization, their analysis shows how the promise of individual movements (e.g., Fair Trade) need not be analyzed in isolation. Likewise, different forms of contemporary social action can be conceived of as unique tactics, each of which plays a role within a broader "war of position."

Finally, my approach follows Evans, who defines similar contemporary social movements as counterhegemonic because they are posing a challenge to "business as usual," rather than because they are likely to "overturn the whole apparatus" (Evans 2000, 231).

Yet, even though social action is manifested in numerous ways, for different purposes, and by a wide range of social actors, it is not clear that it should be celebrated as necessarily contributing to positive social change. For instance, the inclusion of powerful corporations (like Starbucks and Safeway) into Fair Trade commodity chains and the "mainstreaming" of Fair Trade (James 2000) cast doubt on the movement's progressive flair. While seeking to increase sales and find markets for all of the potential Fair Trade producers, Fair Trade organizations are also faced with the task of satisfying "businessmen [sic] who participate in the network and who do so, not from any ideological conviction, but because it is convenient and profitable" (Renard 2003, 92). This encourages observers to question how well Fair Trade networks are able to balance the competing demands that stem from their activist roots and market realities (see also Raynolds 2002).

Thus, if not all social action holds the potential to foster positive change, how can we derive meaningful insights about the theoretical and practical potentials of contemporary expressions of agency? Or, as Starr (2000, 35) asks, shouldn't there be a way to tell the difference between "fashion and a fight"? To answer the questions about the counterhegemonic potentials of the Fair Trade movement posed above, I constructed a typology (see Table 5.1) of forms of counterhegemonic social action in which a contemporary movement working toward social justice in the global agrifood system might engage. These ideal types of social action provide a way to compare and contrast the various manifestations of the Fair Trade banana initiative, to see whether, and if so, how, it challenges the conventional agrifood system and to evaluate its broader potential for fostering progressive and transformative social change in the system.

The typology is based on a conceptual definition of counterhegemonic social action as action undertaken by a social actor or actors, either implicitly or explicitly, in contestation of the hegemony of neoliberal globalization and with the intention of bringing about progressive social change in society. The definition thus encompasses a broad spectrum of actions but excludes those that do not provoke critique of the dominant ideology or contest the status quo. In addition, according to this definition, an actor is presumed to believe, minimally, that an alternative vision of society is needed in order to

Table 5.1   Three Forms of Counterhegemonic Social Action

Form of Action		Possible Implications
• acts of resistance	→	• nonparticipation or partial refusal to participate in the hegemonic system, determined by the actor to be unacceptable • an explicit expression of nonparticipation and challenges to the working of the hegemonic system
• redistributive action	→	• reform of the system to redistribute resources to the benefit of less powerful and/or disadvantaged members of society • redistribution might be seen as a step toward more transformative change that must be fought for in the future
• radical social action	→	• positive, structural transformation of the system resulting in something qualitatively different • transformed system significantly alters patterns of inequality and injustice in prevailing system to the advantage of those previously disadvantaged

really improve it. Finally, the definition implies that any real alternative (i.e., counterhegemony) will involve radical transformation of society.

To further specify what such action looks like, I identify three forms of counterhegemonic action: acts of resistance, redistributive action, and radical social action. They should not be understood as mutually exclusive but can be thought of as hierarchically ordered in terms of their capacity for bringing about structural transformation in society. As summarized in Table 5.1, each type of action is characterized by the implications it has for fostering social change.

Classifying social action according to its transformative potential has a long history. Indeed, a desire to create conditions that will foster social and political change has long captured the attention of social scientists who have, in varying ways, drawn on Marx's anticipation of that moment in which the proletariat transforms itself from a "class of itself" into a "class for itself" (Marx and Engels 1988). For instance, this insight has guided many theorists who have subsequently sought to distinguish between antagonistic and nonantagonistic contradictions of capital (Tse-Tung 1967), revolutionary reforms, and reformist reforms (Gorz 1973), or passive revolution and wars of position and movement (Gramsci 1971). More recently, scholars working on community development and organizing (outside the Marxist tradition) have raised similar concerns. Kennedy and Tilly (1990) suggest a useful distinction between redistributive and transformative populism, while Starr (2000) proposes to differentiate between resistance and struggle.

The conceptual framework I use integrates insights from a diverse group of authors (especially Gramsci 1971; Scott 1990; Kennedy and Tilly 1990; Escobar 1995; Freire 2000; Starr 2000) who have been concerned, in unique ways, with these kinds of questions. The first type, "acts of resistance," draws largely on Scott's (1990) discussion of "infrapolitics" and resistance and refers to action that contests the social system, or an aspect of the social system, that the actor rejects. This type of action is characterized most broadly as nonparticipation in some aspect of the dominant system but does not necessarily involve a rejection of the system entirely. Since hegemony relies on consent, and hence the participation of subordinate and oppressed groups (Gramsci 1971), nonparticipation can therefore be understood as counterhegemonic.

The term "redistributive action" is used to emphasize a primary concern of these actors for making less inequitable the distribution of resources among members of society. In addition to being an expression of resistance, redistributive action also involves a conscious effort to redistribute material resources to those members of a community or society who are disadvantaged by the present pattern of distribution (Kennedy and Tilly 1990). Actors identify the contested system as the source of inequity and/or injustice that is in need of revision. It is counterhegemonic in that it challenges the dominant ideology that tends to hold disadvantaged individuals responsible for their position in society. Although some individuals engaged in this form of action might possess an alternative vision for redesigning the system more radically, a redistributive strategy focuses on bringing about progressive reform to the system.

Finally, I conceptualize "radical social action" as action that is directed toward transforming society as an end goal, explicit in its rejection of the status quo, and reflective of actors' understanding that the sources of inequity and injustice are structural. Actors engaged in this form of action do not necessarily reject reformist strategies, but there is recognition that such initiatives are best understood as partial or short-term solutions. Of these three types of action, this one holds the greatest counterhegemonic potential because, if successful, radical social action will bring about social change that truly transforms the contested system into something qualitatively different. The remainder of the chapter uses examples and insights about the Fair Trade banana initiatives to demonstrate how Fair Trade alternatives have the potential to support action that is at once resistance, redistributive, and radical, and discusses the limits of each of these strategies.

Fair Trade as Resistance

Fair Trade initiatives provide a way for consumers and producers to continue their consumption and production activities but at least partially avoid participating in the conventional trading system. The steady growth of Fair Trade in terms of sales, number of producers involved, diversity of products, and increasing recognition by the public, demonstrates the success of the Fair Trade movement in resisting the hegemony of the conventional agrifood system. Actors are engaged in acts of resistance in the name of Fair Trade in both the global North and the global South.

In the North, socially conscious consumers are, in a sense, voting with their dollars, thereby translating their shopping preferences into a form of resistance (James 2002). This willingness by consumers to purchase food products like coffee, tea, or bananas because of their social significance suggests that the Fair Trade market represents a challenge to neoliberalism and an alternative to competition based solely on price (Renard 1999).

Fair Trade consumption is rising steadily in Europe, where aggregate Fair Trade sales are growing at an average annual rate of 5 percent (EFTA 1998, 25). During the 1990s, retail turnover of some Fair Trade products increased by nearly 100 percent. For bananas in particular, sales have risen by more than 25 percent per year since 1999 (FLO 2002). This consumer activism compares favorably with "acts of resistance," insofar as each Fair Trade purchase represents nonparticipation in the conventional market.

Decisions by important political actors to support Fair Trade can also be interpreted as acts of resistance. The recognition of Fair Trade movement activities by the European Union (EU) is particularly notable. For example, within the EU there was a call to incorporate Fair Trade into its aid policies (European Commission 1999; EFTA 2001a). One member of the European Commission explains the EU's position as follows: "Fair Trade plays a very important role in EU development policies by helping to educate people in the North and by assisting marginalized producers and communities in the South to participate in a more equitable way in the world economy" (remarks by P. Nielson in FLO n.d.). In addition, although this is perhaps more symbolic than anything else, several bodies within the EU have committed to serving only Fair Trade–certified coffee at their meetings (EFTA 2001a, 15). This too reflects the success of the work of Fair Trade organizations and advocacy groups in bringing attention to their cause.

Considered together, these examples support assertions that the Fair Trade movement has achieved important successes in the realm of consumer politics (Raynolds 2000). They also confirm that alternatives to current practices are not only possible but practical and successful (Tiffen 1999), thereby refuting empirically arguments that conventional patterns in the agrifood system are inevitable and demonstrating, as Whatmore and Thorne (1997) argue, that global networking and reach are not unique to transnational corporations.

In the South, Fair Trade–certified producers can also be seen as resisting by their nonparticipation in conventional export channels. However, banana producers who participate in Fair Trade initiatives appear to do so for different reasons than do consumers (see Lyon 2003). This study of small-scale banana production in the Dominican Republic revealed that access to export markets was one of the principal benefits of Fair Trade certification, even if the producers themselves did not always recognize this. For the most part, the producers interviewed for this study shared an unambiguous commitment to supplying any export market, making the increased access to markets an important incentive to participate in the Fair Trade initiative (Shreck 2002b).

There was a very limited level of understanding of Fair Trade, however, even among banana producers from certified Fair Trade associations (see Tallontire 2000). Although slightly more than three-quarters of the producers interviewed were listed as members of Fair Trade–certified associations, only half identified themselves as Fair Trade farmers. When asked more specifically about the Fair Trade initiative (about its benefits and how it worked), even these producers demonstrated only an elementary and partial understanding at best. When asked about Fair Trade, producers with some knowledge often mentioned that they knew there was something called "comercio justo" (Fair Trade), but were unable to explain more. One producer's response captured a sentiment I heard frequently: "Max Havelaar [the name of the Fair Trade organization that worked with these farmers originally] is a guy from Europe and he likes us small farmers, and so he buys our bananas." Others reported more concisely that Fair Trade was "a market" but could not elaborate. Significantly, none of the producers in my sample mentioned or knew about the minimum prices guaranteed by FLO, nor were they aware of the long-term commitment Fair Trade partners are expected to make to producers. By contrast, all of the farmers identified themselves, accurately, as organic banana farmers, revealing a stronger understanding of what organic production was than of what "Fair Trade partner" meant. Given this low level of understanding, can their participation in Fair Trade really be called resistance?

Whether or not producers understand the way the Fair Trade market operates, by selling their fruit through alternative channels they are nonetheless resisting participation in the more exploitive ones. And, like any grower with a harvest to sell, Fair Trade farmers welcome favorable terms of trade that are likely to cover the cost of production. Moreover, as Renard observes, producers who rarely have "the luxury of purist positions . . . are more preoccupied with the struggle for survival and the possibility of increasing sales volumes" (2003, 92). Thus, insofar as survival under oppressive and exploitive conditions is in and of itself an act of resistance (Reagon 1982; Scott 1990), so too is participation in Fair Trade an act of resistance, and it should be understood as counterhegemonic.[6]

The purchasing and producing of Fair Trade commodities should therefore be considered counterhegemonic acts of resistance. However, the locus of this action is centered around the Fair Trade market, which is a critical but contradictory aspect of the Fair Trade model. Although these acts are

6. Moreover, since consumers of Fair Trade products are not expected to reject the capitalist trading system that delivers their tropical commodities to them, neither should the producers be held to such an expectation.

successfully moving Fair Trade from a marginal niche to the mainstream market, several factors limit the potential of this strategy to bring about lasting social change.

First, the structure of international trade (as governed by the WTO and free trade agreements) within which Fair Trade initiatives operate is not necessarily favorable to the continuous growth of the Fair Trade market. For instance, differentiation of commodities according to how they are produced contradicts the WTO's mission of eliminating barriers to trade. Therefore, explicit commitments to supporting Fair Trade efforts are likely to be found unacceptable by the WTO.[7] Recent rulings in the ten-year "banana war" threaten to erode future opportunities for Fair Trade banana farmers in regions where such support is critically needed. Another barrier to market-based resistance stems from the same enthusiasm that contributes to the growth of alternative trade in the first place. Research suggests that consumers and retailers are beginning to suffer from "label fatigue," as the multiplication of competing certification schemes becomes overwhelming and the differentiation between labels becomes confusing and even questionable (Watkins 1998; FAO 2000). A final limitation of this form of resistance is the producer's weak understanding of producers in the Fair Trade market, the initiative more generally, and their role as Fair Trade "partners."

Fair Trade as Redistributive Action

The strength of the Fair Trade movement's counterhegemonic activity stems from its success in implementing a system of certification and labeling to help connect producers and consumers and redistribute resources more equitably. As discussed in more detail elsewhere (Shreck 2002b), the Fair Trade banana initiative brings participating producers much-needed material benefits, improved access to northern markets, and resources for the organizational capacity building of producer associations. In this way, Fair Trade networks serve as an excellent tool for redistributing wealth from northern consumers to southern producers. After the first three years of its banana program, FLO estimated that an average of $2 million was being transferred annually to banana producers in eleven registered associations (FLO 2000b). This money

---

7. Thus far, voluntary decisions regarding Fair Trade are not being challenged by the WTO, but any allowances that could be interpreted as "preferential treatment" would not be permitted. Meanwhile, other WTO decisions undermine possibilities for building up Fair Trade markets, as the following example demonstrates.

comes in the form of higher export prices and a $1.75 social premium paid per forty-pound box of bananas sold with a Fair Trade label.

In the Azua Valley, bananas are regarded as the best of all competing agricultural alternatives. At the time of my research, there were three main banana producers' associations, two of which were certified and registered to sell Fair Trade bananas. As I was told numerous times by producers and exporters alike, when bananas are "good," there is nothing that pays the producers better or as consistently (thanks to the possibility of year-round, biweekly harvests). According to the exporters in the region, with increasing national and international competition the Fair Trade market is the only reason many farmers are able to continue harvesting bananas at all. Data collected on farmers' exporting patterns support this belief. Fair Trade producers were 21 percent more likely to be "currently" exporting (i.e., they had successfully sold their harvest for export at least once within the past month) than were non–Fair Trade producers. Further, more than half (52 percent) of the non–Fair Trade producers, versus only 1 percent of the Fair Trade producers, had been unable to export for more than a year. Thus Fair Trade sales must be understood as having a significant impact within this context.

The Fair Trade model also offers an alternative for consumers who want to partake in a "fairer" form of international trade that does a better job of redistributing resources around the globe. Socially and environmentally conscious consumers are willing to pay higher prices for Fair Trade–labeled commodities in the understanding that they will be helping to improve a flawed system of international exchange. This model forms the core of the movement's strategy of working "in and against" the capitalist market and exemplifies an explicit effort to bring about reform in the agrifood system. This commitment to working within the contested system is critical for the success achieved thus far, as it permits Fair Trade goods to be traded through already established channels without relying fully on conventional players. In addition, it facilitates the entrance of these products into mainstream retail outlets (like major supermarket chains), thereby permitting the largest possible number of consumers to encounter them during their usual shopping routines. Thus, in facilitating the transfer of material resources so that producers receive a larger share of the final price of their product, Fair Trade initiatives hold and realize important potential for redistributive social action.

Nonetheless, the banana initiative's counterhegemonic potential is limited by this redistributive strategy in at least three ways. Situating the Fair Trade banana market within the conventional one creates a situation in which the most important, if unofficial, Fair Trade certification criterion is quality.

Rejection of fruit for failing to meet "exportable-quality" requirements, especially for cosmetic reasons, is so pervasive that it creates a significant structural barrier to the redistributive potential of the initiative. I often documented rejection rates at packing sheds of more than 50 percent. "Exportable-quality" bananas must meet twenty-five criteria if they are to be accepted by the more powerful actors at the other end of the banana chain. In effect, these quality standards institutionalize unequal power relationships between different actors within a commodity chain (Tanaka and Busch 2003), and remind producers that even in the Fair Trade system they are still the least powerful actors in the chain. In addition, many of the criteria are open to interpretation and directly reflect the demands that come with working within a mainstream market, where cosmetic quality is of critical importance. As an FLO bulletin explains, "all farmers know how important quality is. But for banana farmers, shipping across the oceans a highly perishable fresh fruit, it is a weekly headache made worse by the fact that there are no legally binding quality standards" (FLO 2000a, 4).

There are other challenges to redistributing resources once they have been transferred from North to South. The actual redistribution of material benefits within producer communities is hampered by associations' weak organizational capacity, their limited understanding about premiums and minimum prices, and the limited participation of the larger community in making decisions about Fair Trade resources. Low levels of understanding of the Fair Trade system, combined with the relatively young and inexperienced producer associations, mean that most decisions about how to allocate material benefits are made by a small group of leaders in the organizations (Shreck 2002b).

In addition, there is often a lack of price transparency between producers and some exporters, creating confusion among producers about how exporters determine prices. In practice, the Fair Trade minimum prices are received by exporters, who are then expected to pass the benefits on to producers. Yet poor communication about the arrangement and low farmgate prices (the actual price paid to the producers by the exporters) permit some producers to claim that "the company steals from us," or that "there is nothing fair about the price we get." Moreover, fewer than half (45 percent) of the Fair Trade banana producers in the region I studied experienced the security that is expected to accompany an export contract. Each of these challenges poses obstacles to fair and equitable redistribution practices.

Finally, the structure of the banana sector in the Dominican Republic presents additional, though perhaps somewhat unique, limitations, which contribute to the challenges noted above. Fair Trade bananas are better integrated into mainstream distribution channels than are other Fair Trade commodities, like coffee, which is largely differentiated from its conventional counterparts because the fair trading relationships bypass some intermediaries. However, Fair Trade bananas are moved on refrigerated ships that are often controlled by transnational fruit corporations, ripen alongside conventionally traded bananas in special ripening houses, and are then marketed and sold in large supermarket chains. In the Dominican Republic, the Fair Trade associations were unable to bypass even the first intermediary in the chain. Instead, producers (through their associations) sold their fruit to a FLO-certified exporter with whom FLO was working directly and with whom the Fair Trade importers dealt. In other countries, the producer associations themselves are the Fair Trade exporters and receive incentives (like the minimum prices) directly.

Fair Trade as Radical Social Action

What makes Fair Trade unique in the growing sea of certification programs and labeling schemes is that certification is ostensibly awarded on the basis of an alternative trade relationship (Raynolds 2000). This is a crucial distinction in comparison to certification initiatives (such as organic agriculture or the Ethical Trading Initiative) that are based almost solely on the production process. Arguably the most offensive aspects of the conventional global agrifood system do not result from independent production choices made by small-scale farmers in the South.[8] More specifically, Fair Trade organizations seek to make trade relations more transparent, thereby "demystifying global trade and creating more equitable relations of exchange" (Raynolds 2000, 298). Because of this promise, I believe the Fair Trade movement holds an implicit potential to demonstrate the kind of radical social action that confronts and rejects the dominant ideology and practices that uphold the hegemonic system. Up to now, however, the strategies and tactics adopted by Fair Trade actors do not appear to demonstrate this potential empirically.

---

8. For example, writing about ethical trade, Blowfield (1999, 766) cites an example of "independent" decisions. He writes, "But ongoing work with the Ghana pineapple industry shows that part of the reason for the degree of chemical use is to meet consumer demand for a golden fruit, and to ripen fruit at short notice for European wholesalers who seem unable to predict market demand." Many production practices adopted by banana farmers in Azua were similarly guided by "what the company says" to do.

Although not all Fair Trade activists are satisfied with the more conservative goals of reform and redistribution, my research suggests that because of the way Fair Trade initiatives are currently conceptualized and implemented, the implicitly radical and transformative promise of this movement is unlikely to be realized. In considering the limitations to this broader potential, my analysis provides insight into how things like power, participation, and paternalism are articulated in Fair Trade initiatives. It also questions whether the concept of "fair" can be meaningfully transformed into a set of observable criteria by one group for another. Not least, it serves as a reminder of the barriers imposed by Fair Trade initiatives' dependence on powerful actors who pledge allegiance to the capitalist system.

The limitations preventing Fair Trade from conforming to the kind of radical social action defined above can be summarized roughly in three related concerns: the way Fair Trade is being conceptualized by the movement; the top-down implementation of the initiatives; and the concentration of power in the middle of the Fair Trade chain. The Fair Trade model is conceptualized in such a way that it depends on the capitalist market, which presents a fundamental problem for transformative change that would involve rejecting and replacing this contested system. In addition, as part of its strategy, the Fair Trade movement ultimately seeks to facilitate the inclusion of otherwise marginalized groups into the dominant system, which is one way in which it falls short of being truly transformative (Freire 2000). For instance, in order to keep the producer organizations competitive in the Fair Trade market, Fair Trade banana farmers must constantly be educated and encouraged to be better banana exporters. The effects of this conceptualization reverberate in the multiple ways in which Fair Trade is practiced.

Fair Trade initiatives are implemented in a top-down manner. For instance, the groundwork for Fair Trade initiatives was laid by activists who were overwhelmingly from the North. The definition of "fair" and "Fair Trade" has likewise been worked out by Fair Trade organizations based in Europe (EFTA 2001b, 5). Actors representing organizations in the North also decide which producer groups to work with and which commodities to certify. Finally, groups in the North offer producer groups in the South the possibility of being certified and determine the criteria upon which certification is based. This is a central concern for some Fair Trade activists in the South, who also question the unidirectional inspection of southern producer groups by northern Fair Trade organizations and ask who is certifying the North (Gereffi 2000).

Interviews with producers suggest that these kinds of practices make Fair Trade producers feel like aid recipients rather than empowered growers, regardless of what Fair Trade proponents may believe. Producers' lack of knowledge about Fair Trade can be partially explained by this as well; many of them see Fair Trade as yet another scheme dreamed up by foreigners with the stated goal of helping small farmers.

In a related sense, there is a real concern with the distribution of power along the Fair Trade commodity chain. In practice, Fair Trade initiatives are not immune to the demands of powerful actors who control conventional commodity chains. Thus, although FLO is certifying an alternative trade relationship, a closer look at the chain shows that only select exchanges are targeted for alteration. FLO, for instance, "makes no claim to include all actors in the chain" (Lamb and Belling 2000, 43). For banana initiatives, the powerful actors are retailers, ripeners, shippers, and importers. Since most of these conventional players are exempt from complying with the kind of specific Fair Trade criteria that producers are held to, they continue to maintain control in the Fair Trade banana sector. In the end, it seems that the reason Fair Trade organizations must yield to these actors is rooted in how Fair Trade has been conceptualized, relying heavily on the market and its existing distribution channels. This model and strategy reinforce the top-down manner in which the terms of Fair Trade are dictated, largely without the input of producers.

Conclusion

Though sometimes overshadowed by slogans about "equitable partnership," "fair deals," and "trade not aid," Fair Trade organizations are working for sustainable development in the global South. According to the Fairtrade Labelling Organizations' promotional literature, Fair Trade represents a "better deal" for marginalized producers, one that helps put them on the "road towards sustainable development" (FLO 2001). Yet some calls for sustainable development have been criticized for their implicit acceptance of the need for development at all (Lélé 1991; Sachs 1993; Escobar 1995).[9] As Sachs argues,

9. I am not arguing against any form of development in the global South. As critical development scholars have noted, however, the concept of "development" frequently means the path to modernization followed by countries in the North. As history has demonstrated, such "development" was possible in large part because of the North's exploitive relations with the South and with the global environment. This is what makes calls for "development" difficult, as it is neither likely nor desirable that the South should replicate this pattern of development.

"the frame stays the same: 'sustainable development' calls for the conservation of development" (1993, 10). In a similar way, Fair Trade projects can be challenged for calling into question the injustices that characterize the system of free trade by advocating reform rather than overthrow. Thus paternalistic, often neocolonial relations that support current patterns of trade in tropical agricultural commodities are not really challenged. Instead, Fair Trade attempts to make current patterns of exchange more equitable. But colonialism is based on dependency, and even a "fair" version of a dependent relationship can reproduce the same patterns, albeit under a more socially and environmentally friendly guise.[10]

Meanwhile, dependency stretches its reach far beyond the Third World. In the North, consumers are highly dependent on export agriculture, and particularly on the export dependency of producers in the South. Our lifestyles are intimately connected to the very relations that are challenged by the Fair Trade movement. It is here that the contradiction of market dependency expresses itself most sharply. Reliance on the market as the engine of change reinforces the top-down approach to achieving change. Thus Fair Trade perpetuates the North's power to dictate the type and volume of production in the South. Furthermore, reliance on the market (and by extension on consumption) to bring about positive change at the level of production legitimates the primacy of "quality" in Fair Trade exchanges. Waridel and Teitelbaum (1999) explain: "Very few people are ready to spend money on a product that doesn't taste good even if they believe in the cause it represents. Fair Trade products must have the same characteristics as their conventional counterparts especially in terms of taste, and wherever possible, of cost."

This cautionary insight suggests the limits of the counterhegemonic potential of Fair Trade initiatives. Some Fair Trade activists would like to confront the current agrifood system with a more radical challenge. But the commitment to use the capitalist market as the main vehicle for delivering Fair Trade products to consumers may preclude more explicitly oppositional tactics.

In the end, if supermarkets refused to sell Fair Trade bananas, the market would never be able to prosper as it has, and producers would not receive the level of financial support that they do. At the same time, supermarket participation would undoubtedly diminish radically if Fair Trade organizations held retailers to a set of criteria similar to the one with which producers

---

10. I am not advocating the rejection of international exchange and a return to pure localism. On the contrary, I believe it is important for the Fair Trade movement to take seriously the historical dependence of many primary producers in the South on international trade.

must comply. The compromises made for the overall success of the move-ment thus make sense.

There are many winners in the Fair Trade battle of the far larger war of position: consumers who are able to make purchases that satisfy their con-sciences; supermarkets that draw these consumers by supplying these prod-ucts; producers who sell at higher volumes and for better prices; and the Fair Trade movement, which is growing and gaining increased recognition internationally. In the short term, sacrificing more fundamental change (via radical social action) for immediate results seems reasonable. In the longer term, however, can Fair Trade initiatives realize more of their counterhege-monic potential, and in turn alter conventional trading relations? Even if Fair Trade initiatives were implemented according to their ideal, a close look at the model suggests that the practical implications of these projects would remain limited because of how it has been conceptualized. At the same time, if it doesn't work toward more transformative change, the Fair Trade move-ment is likely to reproduce and perpetuate some of the inequities and hier-archical relationships that currently characterize international trade. Even though Fair Trade initiatives seem successful in tilting the balance of power, perhaps making trade less unfair for producers, the movement appears to lack a vision of liberation from the so-called free-market capitalist system.

In conclusion, the positive consequences of the strategic decision to work "in the market" must not be overlooked. The Fair Trade movement clearly has been able to harness some of the advantages of selling in mainstream markets and can redistribute these material rewards to producers. Still, by committing to work "in the market," Fair Trade initiatives embrace a polit-ical economy of food and agriculture that grew out of colonial relations, which today are only partially masked as free trade and globalization. With-out the vision or means for emancipation from the system, might Fair Trade be a postmodern form of opium for the masses?

While I do not attempt to redesign the Fair Trade system, I believe it is possible to identify some of the areas in the model that could be reconsid-ered so that the Fair Trade movement might represent a more radical basis for opposition.[11] More specifically, recognizing the complexities described

---

11. Although the impact on the organization and the initiatives is not yet clear, it is important to point out that the FLO has been engaged in a process of restructuring to increase its credibility as a certification body, improve efficiency within FLO as an institution, and increase transparency toward producers and consumers (FLO 2000b). Significantly, a new internal structure in place since January 2002 incorporates producer organizations, industry, and other stakeholders into FLO governance (FLO 2003).

here, I suggest that a reconceptualization of the Fair Trade model should include a rethinking of at least four tendencies inherent in and limiting to the potential of the Fair Trade banana initiative: the lingering power asymmetry between the North and South; a conservative understanding of empowerment by the movement; the limited participation of southern partners; and the unequal distribution of responsibilities along the Fair Trade commodity chain. By addressing these areas, the movement could realize more of its counterhegemonic potential and contribute to longer-term and more radical changes in the future.

REFERENCES

Allen, Patricia, Margaret FitzSimmons, Michael Goodman, and Keith Warner. 2003. "Shifting Plates in the Agrifood Landscape: The Tectonics of Alternative Agrifood Initiatives in California." *Journal of Rural Studies* 19 (1): 61–75.

Blowfield, Mick. 1999. "Ethical Trade: A Review of Developments and Issues." *Third World Quarterly* 20 (4): 753–70.

Brown, Michael Barratt. 1993. *Fair Trade: Reform and Realities in the International Trading System.* London: Zed Books.

Chin, Christine B. N., and James H. Mittelman. 2000. "Conceptualizing Resistance to Globalization." In *Globalization and the Politics of Resistance*, ed. Barry K. Gills, 29–45. New York: St. Martin's Press.

Conroy, Michael E. 2001. "Can Advocacy-Led Certification Systems Transform Global Corporate Practices? Evidence and Some Theory." Working Paper No. 21. Amherst, Mass.: Political Economy Research Institute.

Escobar, Arturo. 1995. *Encountering Development: The Making and Unmaking of the Third World.* Princeton: Princeton University Press.

European Commission. 1999. Communication from the Commission to the Council on Fair Trade. COM (1999) 619. 29 November. Brussels: European Commission. http://europa.eu. int/eur-lex/en/com/cnc/1999/com1999_0619en01.pdf.

European Fair Trade Association (EFTA). 1998. *Fair Trade Yearbook: Towards 2000.* Ghent: Druk in de Weer.

———. 2001a. *Advocacy Newsletter* 1: 3–4.

———. 2001b. *Fair Trade in Europe, 2001.* Maastricht, The Netherlands: EFTA.

Evans, Peter. 2000. "Fighting Marginalization with Transnational Networks: Counterhegemonic Globalization." *Contemporary Sociology* 29 (1): 230–41.

Fairtrade Labelling Organizations International (FLO). 2000a. "News from the Registers." *Information FLO*, April.

———. 2000b. "What About Restructuring?" *Information FLO*, April.

———. 2001. "How Does FLO Work?" www.fairtrade.net/.

———. 2002. "Sales in 2001." *FLO Fairtrade Fruits Newsletter*, April, 1.

———. 2003. "Impact: Facts and Figures." www.fairtrade.net/sites/impact/facts.htm.

———. N.d. *The Impact of Fairtrade Bananas.* Internal document.

Food and Agriculture Organization of the United Nations (FAO). 2000. "Banana Experts Hold Landmark Meeting at FAO." FAO News and Highlights. www.fao.org/.

Freire, Paulo. 2000. *Pedagogy of the Oppressed.* New York: Continuum.

Friedland, William H. 1994. "The New Globalization: The Case of Fresh Produce." In *From Columbus to ConAgra: The Globalization of Agriculture and Food,* ed. Alessandro Bonanno, Lawrence Busch, William H. Friedland, Lourdes Gouveia, and Enzo Mingione, 210–31. Lawrence: University Press of Kansas.

Gereffi, Gary. 2000. "What Is the Long-Term Vision of a Fair Trade Movement?" Report from Fair Trade Workshop, Keystone, Colorado, May.

Gorz, Andre. 1973. *Socialism and Revolution.* New York: Anchor Books.

Gramsci, Antonio. 1971. *Selections from the Prison Notebooks.* Ed. and trans. Quintin Hoare and Geoffrey Nowell Smith. New York: International Publishers.

Grey, Mark A. 2000. "The Industrial Food Stream and Its Alternatives in the United States: An Introduction." *Human Organization* 59 (2): 143–50.

Heffernan, William D. 2000. "Concentration of Ownership and Control in Agriculture." In *Hungry for Profit: The Agribusiness Threat to Farmers, Food, and the Environment,* ed. Fred Magdoff, John Bellamy Foster, and Frederick H. Buttel, 61–76. New York: Monthly Review Press.

Held, David, Anthony McGrew, David Goldblatt, and Jonathan Perraton. 1999. *Global Transformations: Politics, Economics, and Culture.* Stanford: Stanford University Press.

Hinrichs, C. Clare. 2003. "The Practice and Politics of Food System Localization." *Journal of Rural Studies* 19 (1): 33–45.

James, Deborah. 2000. "Justice and Java: Coffee in a Fair Trade Market." *NACLA Report on the Americas* 34 (2): 11–14.

———. 2002. "Consumer Activism and Corporate Accountability." *Journal of Research for Consumers* 3. http://jrc.bpm.ecu.edu.au/index.asp.

Johnston, Josée. 2003. "Counter-Hegemony or Bourgeois Piggery? Food Politics and the Case of FoodShare." Paper presented at the RC-40 miniconference Resistance and Agency in Contemporary Agriculture and Food: Empirical Cases and New Theories, Austin, Texas, 13–14 June.

Kennedy, Marie, and Chris Tilly. 1990. "Transformative Populism and the Development of a Community of Color." In *Dilemmas of Activism: Class, Community, and the Politics of Local Mobilization,* ed. J. M. Kling and P. S. Posner, 302–24. Philadelphia: Temple University Press.

Lamb, Harriet, and Rolf Belling. 2000. "Review of the Implementation of the Fair Trade Certification Programme in Banana Production and Trade." In *Report: Ad-hoc Expert Meeting on Socially and Environmentally Responsible Banana Production and Trade,* 42–43. Rome: Food and Agriculture Organization of the United Nations.

Lélé, Sharachchandra M. 1991. "Sustainable Development: A Critical Review." *World Development* 19 (6): 607–21.

Lyon, Sarah. 2003. "Fantasies of Social Justice and Equality: Market Relations and the Future of Fair Trade." Paper presented at the Twenty-fourth International Congress of the Latin American Studies Association, Dallas, Texas, 28 March.

Magdoff, Fred, John Bellamy Foster, and Frederick H. Buttel, eds. 2000. *Hungry for Profit: The Agribusiness Threat to Farmers, Food, and the Environment.* New York: Monthly Review Press.

Marsden, Terry. 2000. "Food Matters and the Matter of Food: Towards a New Food Governance?" *Sociologia Ruralis* 40 (1): 20–29.

Marx, Karl, and Friedrich Engels. 1988. *The Communist Manifesto.* Trans. Martin Milligan. Buffalo: Prometheus Books.

McMichael, Philip. 2000a. "Global Food Politics." In *Hungry for Profit: The Agribusiness Threat to Farmers, Food, and the Environment,* ed. Fred Magdoff, John Bellamy Foster, and Frederick H. Buttel, 125–43. New York: Monthly Review Press.

———. 2000b. "The Power of Food." *Agriculture and Human Values* 17 (1): 21–33.

Murray, Douglas L., and Laura T. Raynolds. 2000. "Alternative Trade in Bananas: Obstacles and Opportunities for Progressive Social Change in the Global Economy." *Agriculture and Human Values* 17 (1): 65–74.

Murray, Douglas L., Laura T. Raynolds, and P. L. Taylor. 2003. *One Cup at a Time: Poverty Alleviation and Fair Trade Coffee in Latin America.* Fort Collins, Colo.: Fair Trade Research Group, Colorado State University.

Raynolds, Laura T. 2000. "Re-embedding Global Agriculture: The International Organic and Fair Trade Movements." *Agriculture and Human Values* 17 (3): 297–309.

———. 2002. "Consumer/Producer Links in Fair Trade Coffee Networks." *Sociologia Ruralis* 42 (4): 404–24.

Reagon, Bernice Johnson. 1982. "My Black Mothers and Sisters, or On Beginning a Cultural Autobiography." *Feminist Studies* 8 (spring): 81–96.

Renard, Marie Christine. 1999. "The Interstices of Globalization: The Example of Fair Coffee." *Sociologia Ruralis* 39 (4): 484–500.

———. 2003. "Fair Trade: Quality, Market, and Conventions." *Journal of Rural Studies* 19 (1): 87–96.

Sachs, Wolfgang. 1993. "Global Ecology in the Shadow of 'Development.'" In *Global Ecology: A New Arena of Political Conflict,* ed. Wolfgang Sachs, 3–21. London: Zed Books.

Scott, James C. 1990. *Domination and the Arts of Resistance: Hidden Transcripts.* New Haven: Yale University Press.

Shreck, Aimee. 2002a. "Just Bananas? A Fair Trade Alternative for Small-Scale Producers in the Dominican Republic." PhD diss., Colorado State University, Department of Sociology.

———. 2002b. "Just Bananas? Fair Trade Banana Production in the Dominican Republic." *International Journal of Sociology of Agriculture and Food* 10 (2): 11–21.

Starr, Amory. 2000. *Naming the Enemy: Anti-Corporate Movements Confront Globalization.* London: Zed Books.

Tallontire, Anne. 2000. "Partnerships in Fair Trade: Reflections from a Case Study of Café Direct." *Development in Practice* 10 (2): 166–77.

Tanaka, Keiko, and Lawrence Busch. 2003. "Standardization as a Means for Globalizing a Commodity: The Case of Rapeseed." *Rural Sociology* 68 (1): 22–45.

Tiffen, Pauline. 1999. "The Way Forward: How and When the Alternative Traders Can Say That 'the Market Is Wrong.'" Paper presented at International Fair Trade Association (IFAT) Conference on the Business of Fair Trade: Livelihoods, Markets, and Sustainability, Milan, Italy, 9–14 May.

Tse-Tung, Mao. 1967. "On Contradiction." In *Selected Works of Mao Tse-Tung,* 1:311–47. Peking: Foreign Languages Press.

Waridel, Laura, and Sara Teitelbaum. 1999. *Fair Trade: Contributing to Equitable Commerce.* Quebec: ÉquiTerre. www.equiterr.qc.ca/english/coffee/outils_eng/rapport_euro peen/rapport.html.

Watkins, K. 1998. "Green Dream Turns Turtle." *The Guardian* (Manchester), 9 September, 4.

Whatmore, Sarah, and Lorraine Thorne. 1997. "Nourishing Networks: Alternative Geographies of Food." In *Globalising Food: Agrarian Questions and Global Restructuring,* ed. David Goodman and Michael J. Watts, 287–304. London: Routledge.

Zonneveld, Luuk. 2003. "2001–2002: The Year in Review." FLO *News Bulletin,* January. www.fairtrade. net/sites/new/bulletin.htm.

INSTRUCTOR'S RESOURCES

Key Concepts and Terms:
1. Resistance
2. Fair trade or alternative trade
3. Counterhegemony
4. Globalization

Discussion Questions:
1. Compare the significance of Fair Trade from the perspectives of small-scale farmers in the South and Fair Trade organizations in the North.
2. What are the strengths and limitations of a strategy of working "in and against" the market?
3. How does the Fair Trade movement address power inequalities between North and South? Between consumers and producers?

Agriculture, Food, and Environment Video:
1. *Buyer Be Fair: The Promise of Product Certification.* Bullfrog Films, 2006 (57 minutes).

Agriculture, Food, and Environment on the Internet:
1. Fairtrade Labelling Organizations International: www.fairtrade.net/.
2. Global Exchange (see "Fair Trade" section): www.globalexchange.org/.
3. Banana Link: www.bananalink.org.uk/.

Additional Readings:
1. Brown, Michael Barratt. 1993. *Fair Trade: Reform and Realities in the International Trading System.* London: Zed Books.
2. Murray, Douglas L., and Laura T. Raynolds. 2000. "Alternative Trade in Bananas: Obstacles and Opportunities for Progressive Social Change in the Global Economy," *Agriculture and Human Values* 17, no. 1: 65–74.
3. *New Internationalist.* 1999. "Bananas," *New Internationalist* 317 (October) (entire issue).
4. Ransom, David. 2001. *The No-Nonsense Guide to Fair Trade.* London: Verso.

6

SUSTAINING OUTRAGE: CULTURAL CAPITAL, STRATEGIC LOCATION, AND MOTIVATING SENSIBILITIES IN THE U.S. ANTI–GENETIC ENGINEERING MOVEMENT

William A. Munro and Rachel A. Schurman

There is strong evidence that a new social movement against genetic engineering (GE) in agriculture—or the "anti-GE movement," for short[1]—has had a significant impact on the regulation of these new production technologies, the public awareness and acceptance of genetically modified foods, and the economic fortunes of the agricultural biotechnology industry (Barrett 2000; Kilman 2002; Schurman and Munro 2003). Anti-GE activists have catalyzed important new regulatory restraints on the technology, including insect refuge requirements for genetically engineered crops, a multiyear moratorium on new GE crop approvals in Europe, new labeling laws in many countries, and the negotiation of a new biosafety protocol under the auspices of the United Nations. Under activist pressure in the United States, McDonalds backed away from using genetically engineered potatoes to make fries for fear of a consumer backlash, and the supermarket chain Trader Joe's declared itself GE-free. In 1999 a coalition of U.S.-based activists exposed the presence of an unapproved genetically engineered corn (StarLink®) in the food supply, leading to millions of dollars' worth of product recalls, crop "buybacks," and enormous losses for U.S. companies and agricultural exporters (Harl et al. 2000; Lin, Price, and Allen 2001). The Starlink incident also helped galvanize the development of an "identity preservation" system for genetically engineered foods, and added impetus to efforts to establish new international trade standards for GE foods, under the auspices of the Codex Alimentarious Commission.

---

1. We use the term "anti-GE movement" with some misgivings because those in the movement take a range of positions on the technology, from being totally opposed, to wanting to see the technology better studied, regulated, and subjected to democratic discussion. The reader should keep this variation in mind.

The biotechnology industry has admitted that these actions and changes have affected its reputation and hurt it economically. As a result of the activists' multipronged attack, some of the world's largest agricultural biotechnology firms narrowed their GE crop focus to a few select crops and are taking a less "bullish" stance toward the technology (Barrett 2000; Belsie 2000; Bernton 2000; author interviews). The industry was also moved to participate in several stakeholder dialogues, initiated "listening sessions" with ardent critics (Gilbert 2002; Krueger 2001), and formed a well-financed new countermovement, spearheaded by the Biotechnology Industry Organization (Barboza 2000).

These trends offer several avenues for exploring the issues of agency that motivate this volume. Certainly, the emergence of the anti-GE movement depicts the self-willed and purposive actions, relatively independent of structural constraints, that are the markers of agency (see Wright and Middendorf's introduction to this volume). But what kind of agency is it? How can we explain why these activists—a numerically small group by any measure—have managed to incite such significant change? If this is a new social movement, what are its characteristics, and from where does it derive its power? If we are to understand the role of social forces as agents of change—"from below," so to speak—in the agrifood system, it is necessary to answer these questions. This requires us to understand more fully what drives the actors who make up the antibiotech movement, and what their political and organizational strengths (and weaknesses) are. We need, in other words, a social analysis of the movement.

In this chapter we take an in-depth look at the origins and historical development of the anti-GE movement in the United States. To be sure, the anti-GE movement today is a broad transnational movement with roots and bases in many different countries. The U.S. movement is one component of it. But it does not have its origins in a consumer backlash expanding out of Europe in the early 1990s, as some commentators have suggested. Nor has it grown from a radical and disaffected fringe of ecomalcontents. As we show here, the U.S. anti-GE movement has deeper and independent historical roots in a venerable tradition of social movement politics. Yet it also has its own peculiar organizational and ideological characteristics, shaped by the nature of the issue and the historical moment, as well as by the social backgrounds of its core members. Far from being dependent on external developments, it has a longer history of organizing around agricultural biotechnology than any other movement, and it has played an important role in helping to catalyze the emergence of anti-GE movements in other countries, as well as on a global scale (Purdue 2000; Tokar 2001; author interviews). We focus in this chapter on the movement's early history because we believe that this formative phase profoundly shaped the nature of the contemporary movement in the United States, its principal political strategies, and its strengths and weaknesses as an agent of change.

Our analysis is resolutely actor-oriented, and we adopt Long's view (Chapter 3 of this volume) that "actor-defined" issues should be the starting point for elucidating agency.[2] In doing so, we contribute to the discussions of agency and structure advanced in this volume in several ways. First, we focus on the *making* of a collective social actor through a particular kind of "counterwork" (cf. Long's chapter): the collective development of a critical social analysis of the technology and the development model that drives it. Such counterwork reflects the quality of "reflexiveness" that Wright and Middendorf note is a key requirement of agency. It involves the ability of social actors to critically evaluate their environment and to act purposively on that evaluation to (re)shape it according to their will and interests. In this light, we argue that in order to understand the U.S. anti-GE movement, one must start well before the public campaigns aimed at supermarket boycotts and GE food labeling that erupted in the late 1990s. Early activism began in the 1970s, long before agricultural GEOs were actually introduced into the market, and there was no GE *food* to form an issue around. In this context, early activists *made* agricultural biotechnology a social problem, rather than responding to an already emerging public concern. The conditions under which they did so profoundly shaped the character of the emerging movement and the repertoire of political strategies that activists developed to engage their opponents.

In the first place, the movement lacked a mass base. Instead, it emerged around a core group of key activists who drew on a circumscribed set of organizational, intellectual, and cultural resources that they possessed. As we show below, these core movement activists had long histories of engagement with this and related social issues, and a powerful sense of agency associated with their professional and class backgrounds. Many were well versed in the skills and knowledge of organizing, of power-mongering, and of movement adversaries and other relevant actors that comes with years of participating

---

2. Our analysis is based on a combination of historical archival work and more than two dozen in-depth interviews with the activists who formed the heart of the early anti-GE movement in the United States. These interviews were carried out between June 2000 and July 2002. By "core activists" we mean activists who spend most or all of their time on biotechnology-related organizing. Our sample represents a substantial fraction—probably around one-half—of this activist core.

in oppositional politics. They also drew on a distinct but equally salient body of skills and knowledge that derived from advanced technical, scientific, and professional training in disciplines ranging from regional planning to ecology, biology, and physics. Many found a professional home in public interest think tanks and research establishments. These resources, skills, and organizational bases strengthened these activists' "strategic capacity" (Ganz 2000) and made them into effective combatants on the scientific and regulatory battlefields of agricultural technology change. But they also inclined them to engage their political opponents in ways and in institutional settings that limited their public visibility and appeal (e.g., the courts, legislatures, and regulatory agencies).

These characteristics illuminate the dynamic tension between agency and structure that is always present in social action. On the one hand, they underscore Long's point that social relations, understood in terms of both social networks and class characteristics, are essential for agency to become collective, sustained, and socially significant. Over time, anti-GE activists constituted a kind of "critical community" (Rochon 1998) built on strong personal, professional, and intellectual bonds that enabled them to turn science and technology into a fault line of contestation rather than a conveyer belt of authoritative knowledge. On the other hand, the cultural and cognitive embeddedness of these activists in scientific, technical, and professional disciplines pushed them toward a circumscribed set of political strategies that allowed them to engage the issue effectively at a policy level but limited their capacity to engage the public interest and turn the issue into a broader social concern. In this sense, structure can be seen to both enable and constrain social action (see Wright and Middendorf's introduction).

As Long suggests, the constraining effects of structure can be fruitfully explored through the concept of hegemony, which shapes human behavior at institutional, ideological, and cognitive levels by producing "certain shared accounts of the world . . . so firmly embedded within individuals' consciousnesses as to seem to those individuals part of the very texture of their own subjective being" (Crehan 1997, 30). In this chapter, however, we argue that the qualities of "shared accounts" that produce *consent* can equally be applied to the production of *dissent,* and further that these qualities are not only ideological and cognitive but also profoundly normative. It is impossible to adequately explain the sense of community and solidarity that thickened the ties between anti-GE activists and sustained the movement throughout this early period without appreciating the deep, though multifaceted, sense of moral outrage they shared at the course society was taking and the manner in which decisions about this new technology were being made. Though the source of that outrage varied—from corporate greed, to technological over-reach, to the ethics and wisdom of "playing God" with nature—this current of "motivating sensibilities" created a powerful and lasting bond among movement activists. In this sense, a critical feature of the anti-GE movement was its moral dimension, which both generated cohesion in a diverse social movement and strengthened activists' commitment to the issue.

We develop these points below, showing how the social and historical origins of the U.S. anti-GE movement shaped its emergence, its character, and its limitations as an agent of change in the agrifood system. We begin by showing how a number of activist professionals came in the 1970s to focus their activities on a critical social analysis of biotechnology. Two con-textual factors were important to this process. One was the rise of the "new social movements" in the advanced industrialized countries, which provided a sociopolitical and intellectual milieu conducive to the construction of a critical social analysis of technological development. The other was a series of concrete changes in law, institutional context, and political economy that facilitated the emergence of the "life sciences" industry. We then detail the emergence and consolidation of the movement as a collective actor during the 1980s, as well as its organizational and tactical characteristics. Focusing especially on the Biotechnology Working Group, we stress the formation of an expanding activist community molded both by personal relationships and by activists' complementary and interlocking analyses of technology develop-ment. At the same time, however, the professionalized nature of the move-ment pushed activists into a narrow repertoire of actions that limited their ability to generate a strong public profile. This discussion is followed by an analysis of the activists' motivating sensibilities, drawing on interviews with activists to show the deep current of moral outrage that underlies their commitment to the issue and makes them unlikely to give up the struggle. Finally, we suggest that this analysis indicates a movement, resilient and determinedly oppositional, whose capacity to build a social constituency vociferous enough to lead to public rejection of agricultural biotechnology remains questionable. Nevertheless, the movement has made real substantive gains and is likely to remain a significant player on the field of U.S. agricul-tural biotechnology development.

The Roots of Resistance

In order to make sense of the U.S. anti-GE movement, it is necessary to begin with two historical trends, originating in the 1960s and 1970s, that helped to define both the ideological sensibilities and the professional preoccupations of anti-GE activists.[3] One was the emergence of what scholars have called the "new" social movements that arose in the 1960s and 1970s around human rights and citizens' rights issues such as peace, nuclear power, the environment, women's rights, and sustainable agriculture. The other trend involved a number of institutional developments and political-economic changes that were associated with or carried important implications for the development and spread of new technologies. Together, these developments established a formative intellectual, ideological, and organizational milieu that would prove crucial to the subsequent construction of anti-GE activist networks.

"NEW" SOCIAL MOVEMENTS

The "new" social movements of the 1970s were important to the emergence of the anti-GE movement for both organizational and ideological reasons. Organizationally, these movements tended to draw on loose and often overlapping social coalitions organized around particular issues. Their composition tended to be informal, discontinuous, ad hoc, and context sensitive. They involved a shifting miasma of participants, campaigns, networks, and relationships but also produced relatively secure and ongoing not-for-profit organizations (Greenpeace, Consumers Union, Institute for Food and Development Policy). Over time, these movements generated a complex web of activist initiatives, stretching across the Atlantic and Pacific, that were able to link environmental, food and agriculture, and public safety issues, both nationally and globally, with concerns about democratic accountability and choice.

Although these movements mobilized significantly different ideologies, they marked an important, and in some ways novel, engagement with the institutional and organizational features of social power associated with advanced capitalism. As such, they inspired many of the political-economic, moral, and philosophical discourses that became the hallmark of late twentieth-century politics. These movements were the product of post–World War II affluence in Europe and America, which drew on—and elaborated—a complex and variegated critique of late capitalist (or "postmaterial") society. Although they were profoundly informed by normative considerations of social justice, they tended not to be straightforward class-based movements

---

3. From several of our interviews, as well as our reading of Purdue's and others' work on Europe, much of this analysis would appear to hold true for the European anti-GE movement as well.

concerned with the distribution of material goods under capitalism. Rather, they were most concerned with the impact of contemporary social organization on *values* such as autonomy, self-determination, selfhood, and identity (what Jurgen Habermas [1981] calls the "grammar of forms of life") rather than formal rights. According to some observers, these were "quality-of-life" movements as well as expressions of alternative cultural identities (Habermas 1984; Offe 1985).

These new social movements' critique of modern society had two components that are key to understanding the roots of anti-GE activism. First, they were profoundly concerned with the social and environmental costs of postindustrial society. A critical issue was the increasingly obvious capacity of humans to destroy the earth and their own health and well-being with new technologies: the threat of obliteration posed by nuclear power, the scarcity of nonrenewable natural resources, the large industrial interventions in local and global ecosystems, the massive ecosystemic impacts of industrial agriculture. Though these were large, overarching, often "green" problems, they had a very tangible presence in citizens' own experiences of increased pollution (especially toxics and acid rain), the fear of nuclear meltdowns (dramatically symbolized by Chernobyl and Three Mile Island), the crumbling of urban environments and infrastructures, and the ever-present threat of cancer. Second, the displacement of the social and environmental costs of this economic development model to poor communities, both domestically and globally, motivated activists to work on issues related to human rights, environmental justice, appropriate technologies, and sustainable livelihoods/development.

It is striking how many of the first-generation activists and organizations in the anti-GE movement got their start working on these issues in the 1970s. Cary Fowler, one of the early critics of agricultural biotechnology, came to the "seeds" and biodiversity issue from his work on the root causes of world hunger at the Institute for Food and Development Policy in the mid-1970s. Part of his research involved developing a critique of the green revolution and its impact on small-scale farmers in the global South.[4] Jack Doyle, a policy analyst for the Environmental Policy Institute and author of one of the earliest critical books on biotechnology, worked first as an environmental advocate on transportation and energy issues, before following companies such as Royal Dutch Shell and others into the biotechnology sector.[5] Philip

4. The book was the pioneering *Food First: Beyond the Myth of Scarcity,* co-authored with Frances Moore Lappé and Joseph Collins in 1977.

5. Jack Doyle, interview by authors, Washington, D.C., June 2002.

Bereano, who began working on the DNA controversy and genetic engineering technologies in the late 1970s, was actively involved in the anti–Vietnam War movement and the academic-activist organization Science for the People, established in Cambridge, Massachusetts, in 1969. And Martin Teitel, who was highly instrumental in helping to fund anti-GE activism in the 1980s and became the director of the Council on Responsible Genetics (CRG) in the late 1990s, ran war-zone feeding programs for the American Friends Service Committee and worked on the Youth Project for promoting community development in the 1970s before helping set up the CS Fund, a public interest foundation, in 1981. One of the key issues that the fund took up was nuclear power.[6]

Indeed, the antinuclear movement was a central element in the milieu in which many anti-GE activists moved. It was particularly significant because it brought together a variety of sensibilities, ranging from peace and nuclear disarmament to toxic waste, under the umbrella of concerns about a potentially apocalyptic and unnecessary technology. The growth of the antinuclear movement reflected a growing perception that the organization of modern industrial societies was imposing critical limits on the quality of life of ordinary people: What was the point of being materially secure (under the welfare state) if you lived under the constant threat of being destroyed by nuclear fallout, toxic waste, pollution? As a result, these activists became energized by a growing appreciation that, for all the liberating effects of modern society, life in the modern world was hedged with very potent, and quite terrifying, risks and dangers. What is more, these risks and dangers were themselves an outcome of the institutional organization of a technology- and private property–based economic growth model.

It was precisely these conditions that provided the underpinnings for what Ulrich Beck termed "risk society": a situation in which the risks associated with modern life and development choices were seen to surpass the management capacity of social institutions.[7] Given that the risks associated with late modernity were seen to be potentially species-threatening, concerned citizens began to place a high premium on how society would decide what levels and kinds of risks are acceptable. For "new social movements" activists, this raised critical questions about accountability, responsibility, and

6. Martin Teitel, interview by authors, Cambridge, Mass., January 2002.

7. Indeed, a mounting body of evidence suggests that agricultural genetically engineered organisms cannot by controlled by existing regulatory institutions. Examples include the Starlink crisis, the unexpected appearance of genetically engineered corn in Mexico, and the recent case of Prodigene, in which a biopharming experiment in Iowa went awry.

democratic voice/choice. For them, the pressing question was whether the existing technology- and growth-based model of social organization was really the best institutional arrangement for securing the quality of life (and selfhood) of citizens, as well as the future of society. In the words of long-time anti-GE activist Andy Kimbrell, "there was a change, a significant change. . . . What you had was people who began to say, 'Well, wait a minute, we want to question the whole industrial paradigm.' . . . Part and parcel of what we tried to do in the '60s, whether it be war or something else, [was to] not allow other people to create these huge systems where we weren't actually understanding our responsibility."[8]

STRUCTURAL AND INSTITUTIONAL CHANGES

In addition to the general political milieu established by this broad array of new social movements, key structural and institutional developments helped to shape activist concerns in the 1970s and 1980s. Most obvious was the development of biotechnology, led by the private sector. Following the gene-splicing breakthrough of Stanley Cohen and Robert Boyer in 1973, industry interest grew rapidly. The first biotechnology company, Genentech, was formed in 1976, ushering in an era of "scientist-entrepreneurs" who had one foot in academic research and the other in high-tech start-ups (Kenney 1998). Between 1979 and 1983 more than 250 small biotech firms were founded in the United States (Dibner 1986, cited in Fowler et al. 1988, 183).

This rapid development of technology within the private industrial sector caught the attention of citizens and activists who saw technology as inescapably bound up with social relations and who were deeply concerned about what private enterprise—and the state, for that matter—could do with the technology. A chief concern was that genetic engineering would lead to the commodification of life (including human life), exacerbate social and economic inequalities (both in the United States and between the global North and South), enable genetic discrimination at the workplace, and foster a biological arms race.[9] As the authors of an early issue of *Development Dialogue* on biotechnology bluntly put it, "Any new technology introduced into a society which is not fundamentally just will exacerbate the disparities between rich and poor" (Fowler et al. 1988, 25).

8. Andy Kimbrell, interview by authors, Washington D.C., February 2002.
9. The last was of particular concern to the Council for Responsible Genetics, for example.

Concern over corporate control of these new technologies was heightened by the 1980 civil case of *Diamond v. Chakrabarty,* in which the U.S. Supreme Court ruled that genetically engineered microorganisms are legally patentable. In effect, as many industry analysts—and four dissenting judges—realized, this meant that life itself could now be subject to exclusive monopoly patents, so long as it met the standard criteria of patentability: novelty, utility, and nonobviousness. The meaning of the *Chakrabarty* decision was also immediately intuited by a small group of activists who had been watching the case closely, and who viewed the decision as an "enclosure of the commons" and an extension of the capitalist commodification process, albeit in a qualitatively new realm. As Jeremy Rifkin, one of the first and most important activists on the biotech issue, put it, "I sat in the Supreme Court when the hearing was held on *Chakrabarty.* There were only a few of us there, and I knew that that would be the commercial begetting of the next two centuries" (interview by authors). For many philosophically oriented critics, these developments brought into sharp focus not only the problem of humans' relationship with other forms of nature, but also what it means to *be* human.

Fears about the close relationship between technology development and corporate power were readily fueled by the changing structure of the enterprise as the pharmaceutical, biotechnology, agribusiness, food, chemical, and energy sectors integrated increasingly into a major new "life sciences" industry. During the 1980s, corporate and university relations also became much closer, thanks to the Bayh-Dole Act of 1980, which for the first time allowed the commercialization of federally sponsored research in the United States (Boyd 2003; Kenney 1986). The effect of these changes was not only to blur the distinctions among these once highly differentiated sectors but also to motivate a trend toward consolidation within the biotechnology industry, a trend that a handful of critics were watching carefully. By the end of the 1980s, fewer than a dozen large multinational corporations—or "gene giants," as the Rural Advancement Fund International (RAFI) dubbed them—had emerged to dominate the agricultural biotechnology industry, as the process of corporate concentration continued and large pharmaceutical, food, oil, and petrochemical corporations established linkages with hundreds of small biotech "boutiques" and seed companies (Fowler et al. 1988).

These structural economic changes, together with the *Chakrabarty* case and other legal developments,[10] were key factors in motivating people to start

---

10. Among them the 1970 Plant Variety Patent Act (PVPA) and amendments made to the act in 1980, which allowed firms to patent changes made to plants. The PVPA is discussed in detail in Kloppenburg (1988) and Doyle (1985).

organizing seriously around biotechnology. Jack Doyle, author of *Altered Harvest,* described what was occurring as "an economic race to own the biological and genetic ingredients of agriculture," while another group in the movement's *intelligensia* wrote, "Let us state the problem unequivocally: The greatest threat in the new biosciences is that life will become the monopoly property of a few giant companies" (Fowler et al. 1988, 29). "What is at stake," suggested these same critics a few years later, "is the integrity, future and control of the first link in the food chain. How these issues are decided will determine to whom we pray for our daily bread" (Fowler and Mooney 1990, xiv). Jeremy Rifkin noted that his group started work on Monsanto in 1982, knowing that it would be a long struggle (interview by authors).

These changes in the structure of the life sciences industry (particularly the acquisition of seed companies by Shell Oil, Sandoz, Ciba-Geigy, Monsanto, Upjohn, and others) and its legal architecture also generated concerns about the impact that industry concentration and plant patenting would have on farmer livelihoods and biodiversity in the global South. Many analysts working for public interest organizations worried that the deepening concentration of power and control in the hands of a small number of transnational seed companies would accelerate the trend toward the loss of genetic diversity, which had begun with the industrialization of agriculture and the practice of monocropping and had picked up speed during the green revolution. In their book *Shattering,* for example, Cary Fowler and Pat Mooney described the loss of genetic diversity as a "rendezvous with extinction" and "the single biggest environmental catastrophe in human history" (Fowler and Mooney 1990, ix). They were also sensitive to the possibility that the new biotechnologies could lead to crop substitutions and increases in yields that would put poor southern farmers out of business. If genetic engineering enabled biotechnology firms to produce a product such as vanilla in northern laboratories, they reasoned, those who relied on growing, processing, and exporting vanilla in temperate countries would lose their means of making a living. Similarly, if biotechnology was used to dramatically increase yields of cacao and other crops, as the industry promised, prices would surely fall, leaving farmers in Africa, Latin America, and east Asia in economic disaster (see Fowler et al. 1988, 100–101).

These strands of attention to biotechnology among activists intensified and converged in the 1980s, as private-sector development of the technology gained momentum. It was a critical historical moment: at the very time that the life sciences industry was introducing dramatic new elements of technological risk through its product commercialization, which required a regulatory response from government, the government was beginning to loosen its regulatory framework. Powerful elements in the Carter administration argued that regulation was inflationary and should be curbed (Szasz 1994, 24). The Reagan administration ran with this argument, pushing a stringently deregulatory agenda in which the cost of regulation to industry was to be a core consideration in designing regulatory policy (Eisner 1993). In effect, as a whole array of new technologies was exploding into commercial production, the state began to systematically scale back its role as public watchdog. Not surprisingly, then, citizens concerned about democratic accountability, technology development, and their relationships to private corporate power sharpened their scrutiny. Gradually an organizational framework for independent scientific monitoring and anti-GE activism emerged. This framework comprised a relatively small and professional group of scholar-activists, organizers, and public interest advocates working mainly in the nongovernmental and nonprofit sectors. It provided an institutional and organizational home from which many anti-GE activists could pursue their activities.

The Emergence and Consolidation of the Anti-GE Movement

By the mid-1980s, although there was not yet a real *movement* around agricultural biotechnology, a handful of individuals and groups—Jeremy Rifkin and his staff at the Washington-based Foundation on Economic Trends (FOET); Cary Fowler, Hope Shand, and Pat Mooney at the Rural Advancement Foundation International in North Carolina and Canada; Jack Doyle at the Environmental Policy Institute in Washington; and members of the Council on Responsible Genetics—were working on the issue. Within a few years this small group grew into a critical mass, as activists new and old started to converge on the issue from a variety of backgrounds, issue areas, and perspectives (see Figure 6.1). This core group included, among others, representatives from the aforementioned organizations, along with Michael Hansen from the Institute for Consumer Policy Research, Margaret Mellon and Jane Rissler from the National Wildlife Federation,[11] Rebecca Goldburg from the Environmental Defense Fund, Monica Moore from Pesticide Action Network, Eileen Nic from the International Organization of Consumers Unions (IOCU), Nachama Wilker from the Council for Responsible Genetics, Philip

---

11. Both of these women later moved to the Union of Concerned Scientists, where they continue to work on biotechnology and food-related issues.

Bereano from the University of Washington (and a member of the CRG), Howard Lyman from the National Farmers Union, and Chuck Hassebrook from the Center for Rural Affairs in Walthill, Nebraska. Thanks to a small grant from a California-based foundation, many of these individuals came together face to face in the late 1980s and decided to form the Biotechnology Working Group (BWG). In the process of working together over the next few years, this loose network of activists began to constitute itself as a "collective actor" (Melucci 1996).

The BWG played a catalyzing role in bringing these activists' diverse trajectories to converge, both organizationally and tactically, on the issue of agricultural biotechnology. On a fairly regular basis, BWG members would get together to discuss recent developments in the technology and industry, and to brainstorm action strategies. Thus the BWG was an important place

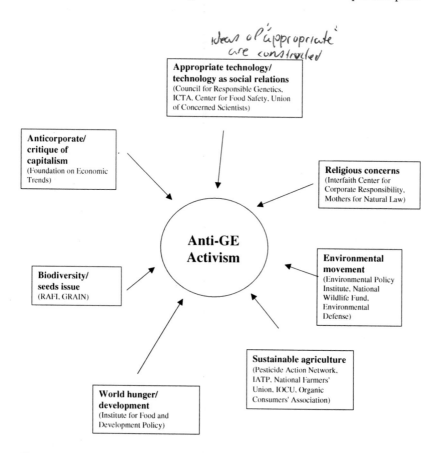

Figure 6.1   Routes to Antibiotech Activism (Exemplary Groups)

for gathering and exchanging information and for forming a systemic political analysis of biotechnology. But, as in most social movements, these meetings provided more than just a forum in which activists could talk, network, and develop a coherent analysis; these face-to-face interactions were crucial in forging the intimate personal relationships and strong sense of commitment, solidarity, and mutual support that helped to sustain the movement, making it hum with energy, tension, humor, and excitement.[12] They were also an important source of inspiration and morale for BWG members. As one member nostalgically recalled, "I have really fond memories [of the BWG] because initially it was really a wonderful group. . . . I mean, I've been to some [other] meetings, and people go, 'oh, this was like the BWG in the old days.'"

Within a few years the BWG had doubled from a dozen or so people to twice that number. Several representatives from family-farm groups joined, as did members of the National Toxics Campaign, the Minnesota Food Association, and the United Methodist Church in Washington, D.C. The inclusion of religious organizations indicated both the range of philosophical and civic concerns raised by the development of the technology and a significant thickening of the social networks involved in the critical social analysis of biotechnology. This was not simply a gathering of policy wonks and scientists, though technical expertise did lie at its core. While virtually everyone had some type of institutional affiliation, usually with a public interest organization, few people got involved at the behest of their organizations; rather, they chose to join the BWG because the issue resonated with them personally. Indeed, some had to expend considerable effort to convince their organizations, which were not that clearly or obviously "connected" to the issue, that they should support their work on agricultural biotechnology.

In 1990 the BWG published a report entitled "Biotechnology's Bitter Harvest: Herbicide Tolerant Crops and the Threat to Sustainable Agriculture," written by Rebecca Goldburg, Jane Rissler, Hope Shand, and Chuck Hassebrook. The report was at once scientifically informed, technically sophisticated, and professional. But it was not only the report that possessed these qualities; these terms went a long way toward describing the anti-GE movement itself during its formative years. Unlike the civil rights, anti–Vietnam

---

12. The importance of these personal relationships was powerfully evident in our interviews, in which most activists referred to each other by first name (as if who they were talking about was self-evident) and were clearly in close, sometimes daily, contact. During our interviews, activists from other organizations would frequently call, send e-mails, or even walk through the door. The abstract term used to describe such interactions, "networking," often causes one to lose sight of the close personal relationships involved.

war, and antitoxics movements, which began as grassroots efforts and eit......
stayed that way or became increasingly professionalized over time as committed activists moved into or created their own organizations (Szasz 1994;
Taylor 1995)[the U.S. anti-GE movement *started out* as a largely professional,
"expertise-oriented" movement.]One reason was that the technology at this
stage had a very low public profile. With no products yet on the market, the
threats that could be posed by agricultural biotechnology remained opaque
and hypothetical in the public consciousness. This was compounded by the
highly technical nature of genetic engineering, which made the issue inac-
cessible to most of the population. As one activist put it, "with this one [i.e.,
biotechnology], you have to talk about life, what does life mean, you have
to start your sentence with 'meiosis' and 'mitosis,' these little quirky things,
getting people comfortable with some basics about genetic science and then
kind of figuring out what that means. There's a lot of [work involved]"
(interview by authors).

This is not to say, however, that the emerging movement was entirely dis-
engaged from grassroots activism. Indeed, some of the earliest mobilization
and organization around biotechnology in the United States was driven by
public safety concerns at the local level about recombinant DNA (rDNA) in
the late 1970s, as universities jumped on the bandwagon of gene-splicing research and began to build specialized laboratories for potentially hazardous
experiments. In similar vein, a small group of concerned citizens in Tulelake
and Monterey counties, California, helped to ignite a major controversy in
the mid-1980s by mounting a significant challenge to an open release field
trial of the "ice minus" bacterium (see Krimsky and Plough 1988; Tokar
2001). The struggle around bovine growth hormone in the late 1980s and
early 1990s also involved some grassroots organizing and resistance. But, for
the most part, the issue of agricultural biotechnology initially attracted the
attention of a small, select group, namely, those engaged in the environmental sciences, where genetic engineering techniques were "hot," those working
on agriculture-related issues (e.g., farm foreclosures, pesticides, sustainable
agriculture), and those concerned about the growing power of corporations.

Strategic Location, Cultural Capital, and Strategies of Contention

By the early 1990s the anti-GE movement had begun to take on a clear shape
and character. As suggested above, the movement showed considerable ideological and organizational continuity with the "new" social movements that

arose in the 1960s and 1970s. Ideologically, anti-GE activism reflected a 1960s counterculture, not of the "turn on, tune in, and drop out" sort but of the sort that involved a powerful element of critical thinking, left-leaning political analysis, and a rejection of the excesses of capitalism and U.S.-style imperialism. Organizationally, the emerging movement consisted of loosely interlocking networks of activists who had come together from a variety of perspectives to work on an issue they shared in common. Many of these individuals, moreover, were located in well-established nonprofit organizations—some of them, ironically, more durable than their corporate adversaries. Others were based in academia. As they came together, the "social movement community" (Buechler 2000) they formed was greater than the sum of its parts.

The fact that many of the people who organized around biotechnology were longtime activists coming from established organizations and institutions, and/or other social movements and struggles, carried two important implications for the nature of the movement. First, it meant that these activists brought with them a wealth of organizing skills, knowledge, and experience, all of which could be brought to bear in the struggle against their corporate and government adversaries (i.e., biotechnology firms and pro-biotechnology government agencies). Years' worth of organizing experience had increased what Marshall Ganz (2000) refers to as activists' "strategic capacity," that is, their ability to interact effectively with their environments and to mount effective strategies, as well as knowledge of their adversaries. Second, it meant that many members of the anti-GE movement operated from organizational bases that were financially and institutionally secure as well as politically engaged. Some came from mass membership organizations, which augmented their ability to educate sizeable constituencies (e.g., the Public Interest Research Groups, or PIRGs, the Sierra Club, Greenpeace, churches), while others came from public interest organizations that had strong scientific and legal capabilities and reputations. Of critical importance was that these organizations had pushed their way into policymaking circles and frequently interacted with regulatory agency officials. Moreover, after more than a decade of Reagan-era deregulation, these environmental, consumer, and other kinds of advocacy organizations had become institutionally entrenched as public watchdogs. From such strategic vantage points, anti-GE activists were well placed to engage in a public policy battle.

Their choice of organizing strategies was profoundly shaped by their particular cache of cultural capital. Not unlike the activists in many other new social movements, white middle-class professionals, many of them with advanced degrees in science, law, economics, or planning, formed the core of the anti-GE movement (Table 6.1). Together, these personal characteristics—class, race, education, and experience—imbued these actors with a strong sense of confidence in their own agency. This was clearly revealed in a comment made by one longtime anti-GE activist who, when asked why he had gotten involved in such a difficult and uphill battle, exclaimed, "*People* make history, either by active involvement or by letting someone else do it for you, and maybe not in your interest. *People make history!*" Armed with good educations, literacy in science and law, and professional credentials, such individuals were not afraid to take on the scientific, legal, and political establishments; in fact, these were the terrains on which they felt most comfortable and where their particular sets of skills and experience could be used to greatest effect. Thus, in a process very similar to the one outlined in Skladany's account of struggles over salmon aquaculture (Chapter 7 in this volume), one of the movement's primary forms of political engagement was the "mobilization of counterexpertise."[13] Much of the work individuals such as Jane Rissler and Margaret Mellon, Rebecca Goldburg, and Michael Hansen carried out, for example, involved going to regulatory hearings; requesting information from the three agencies charged with regulatory oversight of the technology (the U.S. EPA, FDA, and USDA); preparing highly technical critiques of the scientific evidence these agencies were using to make their regulatory decisions; and offering alternative (and far more critical) assessments of the risks posed by genetically modified organisms to the environment and public health. That bona fide scientists were making these arguments bolstered their legitimacy and made their claims harder to discredit. It also enabled the anti-GE movement to provide a powerful counterweight to the industry's exclusively positive portrayal of the technology and its argument that the release of these organisms posed little or no risk to public health or the environment.

Other activists used their backgrounds in law and their knowledge of policy and the legal system to challenge the testing and commercial introduction of genetic engineering technologies. Jeremy Rifkin, Ted Howard, Lee Rogers, Andy Kimbrell, and Joe Mendelson at the Foundation on Economic Trends (and, later, the International Center for Technology Assessment and the Center for Food Safety) brought legal suits and petitions against the U.S. government for its failure to comply with existing laws and policies. One of the first such efforts was an amicus brief filed in the *Chakrabarty* case

---

13. Purdue (2000) makes this same observation about the antigenetics movement in Europe.

Table 6.1   Experience, Educational Level, and Training of First–Generation Antibiotech Activists (Partial List)[1]

Person	Organization	Previous work/activist experience	Professional/ advanced degree	Age (in 2002)
Philip Bereano	Council for Responsible Genetics;[2] Washington Biotechnology Action Council; Professor, University of Washington	Anti–Vietnam War resistance; early member of Science for the People	JD (Columbia University); MA in City and Regional Planning (Cornell University)	early 60s
Jack Doyle	Environmental Policy Institute (later Friends of the Earth)	Environmental issues, transportation at EPI since the mid-70s	MA in City and Regional Planning (Pennsylvania State University)	mid-50s
Cary Fowler	National Sharecroppers' Fund/Rural Advancement Fund International (RAFI)	Worked at Institute For Food and Development Policy ("Food First")	BA (Simon Fraser University, Canada); PhD (University of Uppsala, Sweden)	mid-50s
Rebecca Goldburg	Environmental Defense Fund (now Environmental Defense)		MS, Statistics; PhD, Ecology and Behavioral Biology (University of Minnesota)	mid-40s
Michael Hansen	Consumer Policy Institute (CPI)[3]	Pesticide issues; sustainable agriculture	PhD in Ecology and Evolutionary Biology, (University of Michigan)	late 40s
Andrew Kimbrell	International Center for Technology Assessment (ICTA), Center for Food Safety (CFS)		JD (New York University)	late 40s
Jonathan King	Council for Responsible Genetics (CRG)	Professor of Molecular Biology; antiwar activist; involved with Science For the People	PhD in Molecular Biology (California Institute of Technology)	early 60s

Sheldon Krimsky	CRG	Founding member of the CRG; Professor of Urban and Environmental Policy and Planning	MA in physics (Purdue University) MA, PhD in Philosophy (Boston University)	early 60s
Margaret Mellon	Union of Concerned Scientists	Worked at Environmental Law Institute, then at NWF	PhD, molecular biology, JD (University of Virginia)	mid-50s
Joseph Mendelson III	ICTA, CFS	Greenpeace, US PIRGs, worked on nuclear power issues	JD (George Washington University)	mid-30s
Monica Moore	Pesticide Action Network North America (PANNA)	Co-founder of PANNA; began organizing in the early 1980s	MS in Environmental Sciences, Policy, and Management (UC-Berkeley)	mid-40s
Patrick Mooney	RAFI	Early activism with Canadian development advocacy NGOs; worked for OXFAM-UK.		mid-50s
Jeremy Rifkin	Foundation for Economic Trends	Anti–Vietnam War movement; People's Bicentennial Commission	Business degree, Wharton School (University of Pennsylvania)	late 50s
Jane Rissler	Union of Concerned Scientists	Worked at U.S. EPA and National Wildlife Federation (NWF)	PhD in Plant Pathology (Cornell University)	mid-50s
Hope Shand	RAFI	Worked briefly for ILWU	BA (Duke University); MA in Regional Planning (Population and Development) (University of North Carolina at Chapel Hill)	mid-40s

1. "First-generation" activists are defined as those who began working on the issue in the 1970s or 1980s.

2. Only three of the original members of the Council for Responsible Genetics are included here, mainly for illustrative purposes. In fact, the CRG's founding mothers and fathers included a number of others with PhDs in the natural and social sciences and humanities.

3. Formerly the International Organization of Consumers Unions, or IOCU. CPI is part of Consumers Union.

discussed above, calling attention to the immense and disturbing conse-
quences of a positive Court decision (or what Justice Berger referred to as
Rifkin's "parade of horribles"). Another early initiative was the 1983 lawsuit
Rifkin and others brought against the National Institutes of Health, which
charged that the planned release of "ice minus" bacteria onto open fields in
California represented a violation of the National Environmental Policy Act
(Krimsky and Plough 1988). In the ensuing years, these and other activists
spearheaded dozens of other legal suits in an attempt to slow the develop-
ment and deployment of genetic engineering technologies.

Although these efforts remained largely out of the public eye, they provided an essential organizational foundation for the expansion of strategies
into a more public arena, which would come later. A key example is the Starlink episode in 2000, in which five public interest NGOs collaborated to have
certain nationally sold brands of corn tacos tested for "contamination" by
Starlink GE corn, which was not yet approved for human consumption.
When the tests came back positive, the activists announced a serious regulatory failure that threw the industry into turmoil. Corn trade with foreign
countries was seriously disrupted; the USDA acknowledged that its regulatory
system had broken down and agreed to pay up to $20 million to buy back
Starlink corn seed from seed companies; food manufacturers recalled nearly
three hundred products; and the industry paid out more than $10 million
to farmers in Iowa alone and $9 million to consumers who complained
of allergic reactions to Starlink corn. Perhaps most significantly, the Starlink
incident moved agricultural biotechnology squarely into public view. In imposing significant costs (and potential future costs) on both the industry and
regulatory agencies, this incident demonstrated the ability of the anti-GE activists to convert their particular skills, expertise, and organizational capacities
into effective political action.

Motivating Sensibilities

While the organizational location and cultural capital of the anti-GE movement shaped the spheres in which these activists chose to engage their
adversaries, this alone offers an inadequate account of the *character* of this
movement.[14] Motivating these actors was a profound sense of moral outrage

---

14. All of the quotations used in this section come from the authors' interviews, unless otherwise noted.

THE U.S. ANTI-GENETIC ENGINEERING MOVEMENT   165

about the course society was taking and the manner in which decisions about technological change were being made. As already noted, the roots of this outrage can be traced to the critiques of late capitalist society mounted by the "new" social movements in which many of the movement's core activists had their start, as well as in the ideological variants of 1960s counterculture—ecologism, communitarianism, socialism, anarchism, feminism. Many activists rejected the predominant values of late capitalist society—a reverence for profit and wealth accumulation, the view that "nature" existed purely for human exploitation and consumption, the emotional and physical alienation that had come to characterize people's relationships to each other as well as to nonhuman nature, the problems of an economically dominated political democracy—and embraced alternative value schemes and moral codes.

The moral outrage and normative sensibilities of activists had multiple dimensions; not all were shared by all activists or emphasized by everyone to the same extent. The biggest outrage for many derived from the widely shared perception that corporations had come to wield an unprecedented and dangerous degree of power in society, which they readily and quite typically abused. "If people think it's about food, I don't think so," observed one of the activists we interviewed. "I think it's about the domination of the means of reproduction of genes and the means of development, proteins. This is about 'the corporations are done with zinc and trees' and now, instead of owning the field, instead of enclosing the field, you just own [the] soybeans." Another activist, who has worked on biotechnology for many years, noted:

> We don't have a blanket [opposition] to genetically modified [organisms]. . . . Really it's a matter of control and ownership, and that's why we've always focused on those issues . . . when you push aside all the stuff, and you look at what has really happened in the 130 million acres that have been planted, and the fact that Monsanto's technology accounts for 91 percent of what's been planted worldwide, what else do you have to know? We're talking about control by a single company. There are not really even five gene giants. There's really only one.

For some, the introduction of these new technologies in the context of growing corporate control and economic globalization was especially worrisome. Biotechnology "comes on the scene when corporations have too much power in the world," observed an activist in her midfifties. "Especially in the global environment, where there aren't in fact any governance structures, it's especially dangerous, because it's going to be driven with greed and opportunity." These sensibilities were heightened by the fact that the companies identified as developing and pushing this technology had none too savory reputations stemming from their past as producers of arms and chemicals. As one of our interviewees explained, "I've had cancer, and that, combined with the fact that I think that pesticide companies, like tobacco companies, have been lying for years about the dangers of pesticides—and the fact that they've gotten away for years with promoting these pesticides, and denying their risks—and now, they're admitting these risks of pesticides and saying, 'buy biotech.' You know, *these are the same companies!*"

Many movement activists sought more democratic input into decisions about these literally life-altering technology choices. Sheldon Krimsky, professor of urban and environmental policy and planning at Tufts University, helped form the Council for Responsible Genetics for precisely this reason: "I joined the coalition in '78, because I felt that things were going too fast and we needed to have some brakes on what was going on. And most of the places weren't doing what Cambridge was doing, where we had developed a much more democratic process and citizens were involved and the scientists were now accountable to the chief public health officer, and we had passed the first law in the country." Their chief concern was the role and responsibility of corporations, which were perceived as having sought to develop and introduce these technologies surreptitiously, with no democratic debate or discussion. Indeed, the biggest offense for many lay in the perception that corporations were unilaterally making decisions about technology choices that carried profound implications and repercussions for the rest of society—or, more accurately, for the whole world—without any public participation: "Here you're unleashing a science and a technology, that we have no idea [what it] means to the long-term evolution of the ecology. And it just dumbfounded me. I mean it really did stop me on a gut level and I said, this is wrong. We need moral debate. We need public debate. We need scientific debate. We're completely eliminating a democratic process for discussing this technology." Another interviewee decried the undemocratic nature of government agencies' close relationship with corporate interests: "If you look at [it], you know, where was the informed consent, where was the debate about biotech being [the way to go]? I mean, we all agree that there are big problems in agriculture. . . . But where was the public debate that said, 'let's put all our resources into biotech as a solution to all of these problems in agriculture'?" As a result, viable alternatives such as organic agriculture or public plant breeding were marginalized in development discourses and funding.

These sensibilities about lack of democratic process and choice were connected to a deeply felt anger that the giant life sciences corporations were aggressively pushing for the "right" to patent life, and were being actively supported in this quest by the U.S. government and U.S. courts. The notion that life could or should be patented appalled many activists and confirmed in the clearest terms their analysis of the seamy side of capitalism. Noted one activist, "this is grandiose, in my opinion, in terms of human history. This is the taking of control of the means of reproduction and the means of development. This is biocommerce. This is turning the living world into [a] product. Products, we become products! We become viewed as items of utility." In the eyes of some movement actors, the commodification of life was redolent of a profoundly alienated society, one in which human beings had reached a terrible and unacceptable state of spiritual emptiness and disconnectedness, from each other as well as from nonhuman "nature." When asked what led her to work on biotechnology, for instance, a forty-seven-year-old activist replied,

> It is primarily ecological, but ecological as opposed to environmental, which is to say that the ecological is much more dramatic, as far I'm concerned, in terms of the comprehensiveness. . . . Because I think that our unmediated experiences in the natural world are a very profound way for people, for me anyway, to come to a greater understanding of what the meaning and potential of being human is all about, and as we live in an ever more self-referential world— that is, everything around us is something that humans have created, and we prevent even the *ability* to have the experience of things that weren't created by humans—I think that *that* diminishes human experience very, very deeply. So what motivates me . . . has more to do with how you preserve opportunities for humans to lead meaningful lives than a specific environmental or ecological matter.

In the words of another, "the technological imagination reflected in genetic engineered foods is one that views limits as ontologically evil. It views all limits as evil, and as an environmentalist, as a human being, I find that very, very disturbing."

For these individuals, the view that human and nonhuman life was something that needed to be improved upon—that *could* be improved upon—reflected the moral impoverishment and excessive utilitarianism inherent in late industrial society. "I think this is *the* most powerful technology we are going to confront in this century, and it does, it *will,* ask us what it means to be human," claimed one activist. "To look at every living thing and think of it as just a big Lego set—just a bunch of genes that you can put together in different ways so they can serve our purposes better—that's not the relationship *I* think we should have to nature." It was also a call to action: "Their answer—by them, I mean, the technologists and the corporations—they say, 'no, we're not going to change our technologies so they fit life, we're going to change life so it fits our technology.' That's why I became so involved in biotechnology. That larger image for the last twenty years has been so horrific to me, that larger proposal." Others were motivated by the concern that such interventions could have terrible and irreversible consequences: "It intuitively seemed like you're basically monkeying with nature, and you know, I worked on what happened when you split the atom and global warming stuff. I mean, the unintended consequences, I knew, would be [inevitable]." As another activist noted, "people don't get that *you can't call it back*. . . . It's not like the blanket consequences of what happens when you spray DDT across a bunch of crops. You know, every single transgenic organism is *sui generis,* and you don't know what the implications of that organism proliferating in the world are going to be."

For such individuals, this array of related concerns created a powerful feeling that one *had* to do something to alter the course of history around these new genetic engineering technologies, no matter what the odds were for change or how long it was going to take to achieve it. "We started our [work] with Monsanto in 1982," noted another activist. "It takes a generation. It's going to take another generation before we clean this up." Another suggested that *not* to do anything about the host of moral and ethical issues raised by genetic engineering would be like not standing up when the Nazis came for the Jews:

> I'm not saying that they [the industry and scientists] are going to be successful at this. I'm saying that even as failures, the results could be catastrophic and ultimately change everything as we know it. And that's why I'm fighting cloning of humans and stuff. I don't see—unlike some of my environmentalist friends—I don't see the line between humans, animals, and plants. You know [the line], "I

wasn't a tomato, so I didn't speak up for the tomato; I wasn't a Jew, so I didn't speak up when they came for the Jews. And then they came for me." Well, we *are* speaking up for the tomatoes and the fish and the chickens, and now they *are* coming for us.

This statement indicates the deep sense of moral responsibility many of these individuals felt to publicly question the development of the technology, as well as the broad connections activists made among its different possible applications (i.e., human, plant, and animal).

Activists' sense of responsibility to *do* something was rooted in part in the personal and class backgrounds outlined above. But it was also grounded in a range of ethical and philosophical sensibilities, whether secular or religious. Some were consequentialist, stressing the unknown but potential (and potentially irreversible) threats that the technology posed to human health or the environment. Others were ontological, stressing the view—from either a secular or a religious standpoint—that humans have no *right* to create and *own* new forms of nature. Yet others stressed the logics of "risk overreach" inherent in a capitalist economy. But they all agreed that the stakes in this struggle were very high, not only in terms of their political engagement with the industry and with regulatory agencies but in terms of their broader vision of a "good" and sustainable society. Their actions were impelled by an oppositional consciousness and a worldview that reflected a different set of values from those embraced by the dominant society, with its reverence for technological change, scientific "progress," and the private corporation. Compelled by deeply held convictions and a sense of outrage, they were unlikely to throw in the towel.

Conclusion

In Chapter 2 in this volume, William Friedland develops an empirical typology of social agency in order to assess the potential of different modalities of agency to constitute transformative social action. He maps social agency on two axes: spontaneous→organized and acceptant→resistant, though he also indicates the importance of a third dimension, persistence. His point is to suggest that we might learn something about the transformative potential of particular antihegemonic actions if we situate them on these continua, in real historical contexts. Our account of the rise of the U.S. anti-GE movement would place it in the lower right-hand quadrant of Friedland's typology:

organized, resistant, and persistent. This suggests that we might be optimistic about its prospects. Yet our account also suggests several ways in which we might usefully refine the typology.

In the first place, the *way in which* the U.S. movement developed into a collective actor had an important impact on its character. The key process in building the movement was the collective construction of a sustained, critical social analysis of the technology. Because the initial anti-GE activists possessed crucial scientific and political skills, were able to draw on (and develop) networks of knowledge generation and information sharing, and occupied strategically located organizational bases, they became well-entrenched players on the playing fields of regulatory policy, public health, and scientific risk debates. Much to the industry's and government's chagrin, these activists could not be ignored or wished away by the major proponents of the new genetic engineering technologies, who saw biotechnology as the (only) way of the future. Indeed, the strategic organizational locations that these activists occupied have become *more* significant over the past twenty-five years, as government regulatory agencies whose presumed job it is to defend the public interest have increasingly ceded their role to nongovernmental and private actors.

Second, the *moral* component of the antihegemonic consciousness that motivated these activists was important in defining the movement's character. Indeed, it was perhaps this normative current of "motivating sensibilities" that ensured that these activists would move beyond Friedland's category of "primitive agency," inspiring them to *action* rather than *withdrawal* even when the prospects for slowing the biotechnology juggernaut seemed dismal. The depth of these activists' outrage about the undemocratic nature of decision making around the technology, and their feelings about what these technologies would mean for the human condition, made it unlikely that the core of the anti-GE movement would be readily co-opted or conventionalized.

Thus the defining characteristics of the anti-GE movement gave it the staying power that Verta Taylor has written about in her work on social movements in abeyance (Taylor 1989). Yet staying power does not translate directly into political effectiveness. The key strengths of these activists— their moral outrage, professional backgrounds, and institutional locations— also pushed the movement in strategic directions that limited its capacity to mobilize widespread *popular* support. For example, many of the reports and publications produced by the early activists in the movement, particularly by the scientists, were quite technical and hence accessible only to a relatively small professional readership that included other scientists, academics, and the highly educated portion of the public. Likewise, many of the legal cases the movement pursued were so esoteric that they were unable to capture the popular imagination and hence remained confined to the courts (even though in some cases they forced the government to make some significant regulatory changes).

Another real constraint on some of the scientist-activists involved in the anti-GE movement was their discomfort with taking positions that were, or could be construed as being, "antiscience," or that condemned a whole class of research tools and techniques—such as genomics—that were rapidly becoming widespread in their fields. Most activist-scientists saw considerable nuance and complexity in the issue, and were uncomfortable adopting the kinds of black-and-white positions or sound-bite moral oppositions that often help attract a mass constituency to a cause. Several of the people we interviewed recognized the limits this placed on their organizing strategies but resolutely maintained that they could act only in ways that felt comfortable to them and that would not jeopardize their—or their organizations'—credibility. For many, the decision to engage mainly in a "politics of (counter)expertise" was based on the calculation that this mode of political engagement allowed them to use their skills in the way they believed would be most effective.

These activists' strategic decisions also reflected a realistic reading of the historical and political moment. When they first started to organize in the late 1970s, and even throughout most of the 1980s, no agricultural GEOs had actually been introduced into the market. During an interview, one activist recounted the frustration of trying to organize a rally around the release of the report *Biotechnology's Bitter Harvest* in front of a Californian biotechnology company in the late 1980s, and not being able to interest anyone, including the press, in the rally or the issue. As she saw it, she and her colleagues were just too far ahead of the curve. Furthermore, the broader political and cultural environment in the United States was not one that easily legitimized more confrontational and mass-based tactics, especially if they involved crimes against property. Indeed, the overall "unfriendliness" of the U.S. political-cultural environment was readily recognized by those in the early anti-GE movement who *did* try to turn this into a mass issue (e.g., Jeremy Rifkin) and who failed to rouse the public. This realization only served to reinforce the movement's proclivity to pursue tactics that were relatively "professional" and nonconfrontational in character (some might even say elitist). The ability of the movement to shape public opinion in the United States thus remained very limited.

These limitations indicate that, the movement's resilience notwithstanding, its efficacy as an oppositional actor must be assessed with some caution. Nevertheless, if it is clear today that the anti-GE movement has not forced agricultural biotech into full-scale retreat, it *has* had real substantive effects. One is that the movement has grown more diverse. Since the mid-1990s, in the United States as well as abroad, movement activities have spread outward into public campaigns aimed at supermarket boycotts and GE food labeling, for example, and downward toward the grassroots, for instance in the farmer-to-farmer campaigns against GE crops in the U.S. Midwest, as well as in local GE-food moratoriums and labeling campaigns in the U.S. Northeast and Northwest. Mainstream environmental NGOs such as the Sierra Club and Greenpeace have been drawn into the fray, as have local food safety organizations and the organic farming movement. Indeed, the organic farming movement has drawn considerable energy from the anti-GE movement. Sensibilities such as those that drive Salmon Nation (see the next chapter) have inspired citizens to campaign for GE-free ordinances in cities such as Ann Arbor, Michigan, and Austin, Texas. All of these campaigns have drawn on the foundational intellectual and organizing work done by the activists discussed here. Thus the activists described in this chapter today occupy crucial nodes in what is in fact a broad—and *broadening*—network of organizations, campaigns, and citizen groups that are publicly expressing concerns about a genetically engineered future.

The movement has also had a substantive effect on the terrain of technology development. Even as the cycle of protest that characterized the late 1990s and early 2000s has died down, it has left a heightened public awareness of risk as its legacy. U.S. government regulatory agencies such as the FDA and the USDA have had to respond to that awareness by increasing public hearings and tightening oversight of industry testing procedures. Under pressure from the anti-GE movement, the biotechnology industry has restructured, expended a great deal of energy and money on public relations, and become more sensitive to the concerns of consumers. Given this record and its members' level of commitment, resiliency, and strategic capacity, the anti-GE movement is likely to continue to have a significant effect on the political battlefields of agrifood system change.

REFERENCES

Barboza, David. 2000. "Industry Moves to Defend Biotechnology." *New York Times,* 4 April, C6.

Barrett, Amy. 2000. "Rocky Ground for Monsanto?" *Business Week*, 12 June, 72.

Belsie, Laurent. 2000. "Superior Crops or 'Frankenfood'? Americans Begin to Reconsider Blasé Attitude Toward Genetically Modified Food." *Christian Science Monitor*, 1 March, 73.

Bernton, Hal. 2000. "Hostile Market Spells Blight for Biotech Potatoes." *Seattle Times*, 30 April.

Bonanno, Alessandro. 1994. "Introduction." In *From Columbus to ConAgra: The Globalization of Agriculture and Food*, ed. Alessandro Bonanno, Lawrence Busch, William H. Friedland, Lourdes Gouveia, and Enzo Mingione, 210–31. Lawrence: University Press of Kansas.

Bonanno, Alessandro, Douglas H. Constance, and Heather Lorenz. 2000. "Powers and Limits of Transnational Corporations: The Case of ADM." *Rural Sociology* 65 (3): 440–60.

Boyd, William. 2001. "Making Meat: Science, Technology, and American Poultry Production." *Technology and Culture* 42 (4): 631–64.

———. 2003. "Wonderful Potencies? Deep Structure and the Problem of Monopoly in Agricultural Biotechnology." In *Engineering Trouble: Genetic Engineering and Its Discontents*, ed. Rachel A. Schurman and Dennis Doyle Takahashi Kelso, 24–62. Berkeley and Los Angeles: University of California Press.

Boyd, William, W. Scott Prudham, and Rachel Schurman. 2001. "Industrialization and the Problem of Nature." *Society and Natural Resources* 14 (7): 555–70.

Buechler, Steven M. 1990. *Women's Movements in the United States: Woman Suffrage, Equal Rights, and Beyond*. New Brunswick: Rutgers University Press.

———. 2000. *Social Movements in Advanced Capitalism: The Political Economy and Cultural Construction of Social Activism*. New York: Oxford University Press.

Busch, Lawrence, and Arunas Juska. 1997. "Beyond Political Economy: Actor Networks and the Globalization of Agriculture." *Review of International Political Economy* 4: 688–708.

Busch, Lawrence, and Keiko Tanaka. 1996. "Rites of Passage: Constructing Quality in a Commodity Subsector." *Science, Technology, and Human Values* 21 (January): 3–27.

Charles, Daniel. 2001. *Lords of the Harvest: Biotech, Big Money, and the Future of Food*. Cambridge, Mass.: Perseus Publishing.

Doyle, Jack. 1985. *Altered Harvest: Agriculture, Genetics, and the Fate of the World's Food Supply*. New York: Viking.

Eisner, Michael. 1993. *Regulatory Politics in Transition*. Baltimore: Johns Hopkins University Press.

Fowler, Cary, E. Lachkovics, Pat Mooney, and H. Shand. 1988. *The Laws of Life: Another Development and the New Biotechnologies*. Uppsala, Sweden: Dag Hammarskjöld Foundation.

Fowler, Cary, and Pat Mooney. 1990. *Shattering: Food, Politics, and the Loss of Genetic Diversity*. Tucson: University of Arizona Press

Friedmann, Harriet. 1993. "The Political Economy of Food: A Global Crisis." *New Left Review* 1 (January–February): 29–57.

Ganz, Marshall. 2000. "Resources and Resourcefulness: Strategic Capacity in the Unionization of California Agriculture, 1959–1966." *American Journal of Sociology* 105 (4): 1003–62.

Gilbert, Virginia Baldwin. 2002. "Reaching Out to Change Perceptions: Monsanto Chief

Executive Hendrik Verfaillie Is Described as a Good Listener Who Is Respectful to Others." *Saint Louis Post-Dispatch,* 12 May.

Goodman, David. 1999. "Agro-food Studies in the 'Age of Ecology': Nature, Corporeality, Bio-politics." *Sociologia Ruralis* 39 (1): 17–38.

Goodman, David, and Michael J. Watts. 1994. "Reconfiguring the Rural or Fording the Divide? Capitalist Restructuring and the Global Agrofood System." *Journal of Peasant Studies* 22 (1): 1–49.

Gouveia, Lourdes. 1997. "Reopening Totalities: Venezuela's Restructuring and the Globalisation Debate." In *Globalising Food: Agrarian Questions and Global Restructuring,* ed. David Goodman and Michael J. Watts, 305–23. London: Routledge.

Habermas, Jurgen. 1981. "New Social Movements." *Telos* 49 (fall): 33–37.

———. 1984. *The Theory of Communicative Action.* Boston: Beacon Press.

Harl, Neil E., Roger G. Ginder, Charles R. Hurburgh, and Steve Moline. Forthcoming. *The Starlink Situation.*

Heffernan, William D., and Douglas H. Constance. 1994. "Transnational Corporations and the Globalization of the Food System." In *From Columbus to ConAgra: The Globalization of Agriculture and Food,* ed. Alessandro Bonanno, Lawrence Busch, William H. Friedland, Lourdes Gouveia, and Enzo Mingione, 29–51. Lawrence: University Press of Kansas.

Juska, Arunas, and Lawrence Busch. 1996. "The Production of Knowledge and the Production of Commodities: The Case of Rapeseed Technoscience." *Rural Sociology* 59 (4): 581–97.

Kelso, Dennis Doyle Takahashi. 2003. "Conclusion: Re-creating Democracy." In *Engineering Trouble: Genetic Engineering and Its Discontents,* ed. Rachel A. Schurman and Dennis Doyle Takahashi Kelso, 239–53. Berkeley and Los Angeles: University of California Press.

Kenney, Martin 1986. *Biotechnology: The University-Industrial Complex.* New Haven: Yale University Press.

———. 1998. "Biotechnology and the Creation of a New Economic Space." In *Private Science: Biotechnology and the Rise of the Molecular Sciences,* ed. A Thackray, 131–43. Philadelphia: University of Pennsylvania Press.

Kilman, Scott. 2002. "Monsanto Cuts Profit Outlook amid Latin American Weakness." *Wall Street Journal,* 13 June.

Kloppenburg, Jack R., Jr. 1988. *First the Seed: The Political Economy of Plant Biotechnology, 1492–2000.* Cambridge: Cambridge University Press.

Krimsky, Sheldon, and Alonzo L. Plough. 1988. *Environmental Hazards: Communicating Risks as a Social Process.* Dover, Mass.: Auburn House.

Krueger, Roger W. 2001. "The Public Debate on Agrobiotechnology: A Biotech Company's Perspective." *AgBioForum* 4 (3–4): 209–20.

Lawrence, Geoffrey. 1999. "Agri-Food Restructuring: A Synthesis of Recent Australian Research." *Rural Sociology* 64 (2): 186–202.

LeHeron, Richard, and Michael Roche. 1995. "A 'Fresh' Place in Food's Space." *Area* 27 (1): 23–33.

Levidow, Les. 1995. "Scientizing Security: Agricultural Biotechnology as Clean Surgical Strike." *Social Text* 13 (3): 161–80.

———. 1999. "Britain's Biotechnology Controversy: Elusive Science, Contested Expertise." *New Genetics and Society* 18 (1): 47–64.

————. 2000. "Pollution Metaphors in the UK Biotechnology Controversy." *Science as Culture* 9 (3): 325–51.

————. 2001. "The GM Crops Debate: Utilitarian Bioethics?" *Capitalism, Nature, Socialism* 12 (1): 44–55.

Lin, William, Gregory K. Price, and Edward Allen. 2001. *Starlink: Impacts on the U.S. Corn Market and World Trade.* Washington, D.C.: USDA, Economic Research Service.

Lockie, Stewart, and Simon Kitto. 2000. "Beyond the Farm Gate: Production-Consumption Networks." *Sociologia Ruralis* 40 (1): 3–19.

Marsden, Terry. 1997. "Creating Space for Food: The Distinctiveness of Recent Agrarian Development." In *Globalising Food: Agrarian Questions and Global Restructuring,* ed. David Goodman and Michael J. Watts, 169–91. London: Routledge.

Marsden, Terry, and Alberto Arce. 1995. "Constructing Quality: Emerging Food Networks in the Rural Transition." *Environment and Planning A* 27 (8): 1261–79.

McMichael, Philip, ed. 1994. *The Global Restructuring of Agrofood Systems.* Ithaca: Cornell University Press.

————. 1996. "Globalizaton: Myths and Realities." *Rural Sociology* 61 (1): 25–55.

McMichael, Philip, and David Myhre. 1991. "Global Regulation vs. the Nation-State: Agro-Food Systems and the New Politics of Capital." *Capital and Class* 43 (spring): 83–105.

Melucci, Alberto. 1996. *Challenging Codes: Collective Action in the Information Age.* Cambridge: Cambridge University Press.

Moore, Kelly. 1999. "Political Protest and Institutional Change: The Anti-Vietnam War Movement and American Science." In *How Social Movements Matter,* ed. Marco Guigni, Doug McAdam, and Charles Tilly, 97–118. Minneapolis: University of Minnesota Press.

Offe, Claus. 1985. "New Social Movements: Challenging the Boundaries of Institutional Politics." *Social Research* 52 (4): 817–68.

Pritchard, William N. 1999. "The Emerging Contours of the Third Food Regime: Evidence from Australian Dairy and Wheat Sectors." *Economic Geography* 74: 64–74.

Purdue, Derrick A. 2000. *Anti-genetiX: The Emergence of the Anti-GM Movement.* Aldershot, UK: Ashgate.

Raynolds, Laura, David Myhre, Philip McMichael, Viviana Carro-Figueroa, and Frederick H. Buttel. 1993. "The New Internationalization of Agriculture." *World Development* 21 (7): 1101–21.

Schurman, Rachel A., and William A. Munro. 2003. "Making Biotech History: Social Resistance to Agricultural Biotechnology and the Future of the Biotechnology Industry." In *Engineering Trouble: Genetic Engineering and Its Discontents,* ed. Rachel A. Schurman and Dennis Doyle Takahashi Kelso, 111–29. Berkeley and Los Angeles: University of California Press.

Szasz, Andrew. 1994. *EcoPopulism: Toxic Waste and the Movement for Environmental Justice.* Minneapolis: University of Minnesota Press

Taylor, Bron Raymond. 1995. *Ecological Resistance Movements: The Global Emergence of Radical and Popular Environmentalism.* Albany: State University of New York Press.

Taylor, Verta. 1989. "Social Movement Continuity: The Women's Movement in Abeyance." *American Sociological Review* 54 (5): 761–75.

Tokar, Brian. 2001. "Resisting the Engineering of Life." In *Redesigning Life: The Worldwide Challenge to Genetic Engineering,* ed. Brian Tokar, 320–36. New York: Zed Books.

Whatmore, Sarah. 1994. "Global AgroFood Complexes and the Refashioning of Rural Europe." In *Globalization, Institutions, and Regional Development in Europe,* ed. Ash Amin and Nigel Thrift, 46–67. London: Oxford University Press.

Whatmore, Sarah, and Lorraine Thorne. 1997. "Nourishing Networks: Alternative Geographies of Food." In *Globalising Food: Agrarian Questions and Global Restructuring,* ed. David Goodman and Michael J. Watts, 287–304. London: Routledge.

INSTRUCTOR'S RESOURCES

Key Concepts and Terms:

1. Biotechnology
2. Identity Preservation
3. Transnational Social Movement
4. New Social Movement

Discussion Questions:

1. Who are the actors in the antibiotechnology movement? What are their grievances?
2. To what extent have anti-GE activists shaped the development of the contemporary agrifood system?
3. Describe the organizational and ideological characteristics of the anti-GE movement. How do they differ from the organizational and ideological traits of the Fair Trade movement actors discussed by Shreck in Chapter 5?

Agriculture, Food, and Environment Videos:

1. *Field of Genes.* Bullfrog Films, 1998 (44 minutes).
2. *This Is What Free Trade Looks Like.* Activist Media Project, 2004 (60 minutes).

Agriculture, Food, and Environment on the Internet:

1. Institute for Agriculture and Trade Policy: www.iatp.org/.

Additional Readings:

1. Charles, Daniel. 2001. *Lords of the Harvest.* New York: Perseus Publishing.
2. Kloppenburg, Jack R., Jr. 1988. *First the Seed: The Political Economy of Plant Biotechnology, 1492–2000.* Cambridge: Cambridge University Press.
3. Middendorf, Gerad, Michael Skladany, Elizabeth Ransom, and Lawrence Busch. 2000. "New Agricultural Biotechnologies: The Struggle for Democratic Choice." In *Hungry for Profit: The Agribusiness Threat to Farmers, Food, and the Environment,* ed. F. Magdoff, J. B. Foster, and Frederick H. Buttel, 107–24. New York: Monthly Review Press.
4. Schurman, Rachel A., and Dennis Doyle Takahashi Kelso, eds. 2003. *Engineering Trouble: Biotechnology and Its Discontents.* Berkeley and Los Angeles: University of California Press.
5. Tokar, Brian, ed. 2001. *Redesigning Life? The Worldwide Challenge to Genetic Engineering.* New York: Zed Books.

7

SOCIAL LIFE AND TRANSFORMATION IN
SALMON FISHERIES AND AQUACULTURE

Michael Skladany

Salmon are at the heart of major social changes taking place in the seafood sector. The recent emergence of global production methods based on the industrial farming of salmon has transformed seafood. The up side to this change has been that salmon, known to be an important part of a healthy diet, has been made more readily available to consumers. Yet studies are revealing that abundant heart-healthy salmon in our supermarkets comes with a high price. It has been documented that industrial salmon farming threatens wild salmon, marine life and the oceanic environment, coastal livelihoods, and indigenous cultures, and in some cases raises human health concerns (research has found that farmed salmon contains high levels of polychlorinated biphenyls, or PCBS, a cancer-causing agent). Unlike the far more advanced anti-genetic engineering movement described in the previous chapter, the contemporary struggle over salmon has only become visible in the past five years. Increasingly, lines are being drawn to distinguish wild from farmed salmon.[1] This distinction between wild and farmed salmon provides a useful means of examining the dynamic coextensive interplay between human agency and social structure, which is the objective of this chapter.

Salmon have a complex biological and social history. This history shows the ways in which human agency is subject to the constraints imposed by the structure of salmon (and more broadly seafood) production and consumption. In other words, humans have shaped, and are actively shaping, the transformation of salmon in the contemporary food system. From precolonial

---

1. Most of the seven "wild" salmon species spend the early part of their lives in the controlled conditions of hatcheries. The socially constructed distinction between "wild" and "farmed" salmon pertains to the source of production, with actors, such as fishers, claiming the "wild" designation for marketing advantages.

North America, to the commercialization of salmon fisheries, to the industrial farming of salmon, agency has been exercised in a number of ways. In this chapter, agency and structure are used as coextensive conceptual underpinnings, meaning that humans act through structure and structure acts through humans (Callon and Law 1997). Canada's First Nations, fishers, corporate aquaculturists, policymakers, environmentalists, chefs, consumers, and grant-making marine conservation foundations are all attempting to reshape the structure of salmon production and consumption. It is thus evident that a fundamental tension exists between human agency and structure (see Wright and Middendorf's introduction to this volume) over salmon. The coextensive relationship between agency and structure is expressed through resistance, conflict, and negotiation, but most symbolically through the emergence of the movement for sustainable seafood choices. It is imperative to acknowledge that contests over salmon are key harbingers of a broader sustainable seafood movement.

Groups that oppose the farming of salmon have been able to achieve some early visibility and to affect market conditions by educating chefs and high-end restaurant patrons, as well as the general public, about the problems caused by farming salmon. While attempting to achieve many of the same objectives as the anti–genetic engineering movement, the campaign strategies undertaken by farmed salmon opponents are primarily concentrated on influencing consumer choice in the marketplace and challenging industrial and federal science-based policy. To maintain market share, the salmon industry has responded by moving quickly to voluntarily adopt environmental certification schemes, challenge scholarship that reflects poorly on it, develop media campaigns that deflect negative publicity, and construct an image of their sector as environmentally responsible. Like other contributions to this volume, the story of salmon is a story of human agency and structure. Unlike analyses of land-based agrifood, however, this chapter describes the contours of indigenous salmon fisheries, salmon commercialization, and the rise of the global salmon aquaculture industry.

This chapter identifies important transition points in the transformation of salmon and concludes with a discussion of the future of salmon as seen through the rubric of a growing social movement. My objective is to explore the role of human agency against the structural backdrop of salmon fisheries and, more recently, aquaculture. The case study of salmon demonstrates that there are vast forces at work in restructuring our seafood system that will be with us well into the future.

Indigenous Salmon Fisheries as Survival

Understanding the historical background of salmon is essential, for it contin-ues to frame and influence the current struggles over salmon. Historically, actors have exerted agency at different social-structural and ecological loca-tions. On the one hand, salmon have become commodities in our modern food system. On the other hand, they retain strong cultural ties for indigenous fishers and others who rely on the fish as a vital food source and for economic livelihood. White (1995) captures the key elements in the transformation of salmon and, by extension, society and nature, when he writes that "preserving salmon is as much a social and cultural matter as a biological and economic one. Throughout modern history, humans have struggled to turn space into property and salmon into a commodity, but this was only part of the trans-formation of nature taking place. . . . In their dying salmon revealed constel-lations of competing social values, and understanding the fate of salmon involves understanding complicated and particular social struggles" (43).

The embeddedness of salmon in the cultures of indigenous people sug-gests that salmon fisheries can provide an initial point of departure for chart-ing these complicated social struggles. Along the northern Pacific coast, the struggle for sovereignty is central to the collective life of indigenous people. Prior to colonization, salmon harvest was a finely tuned cultural activity integrally wedded to the cycles of nature. The arrival and harvest of salmon were marked by a cosmological uncertainty between fish and people. Salmon were crucial to the diets of indigenous coastal people, ensuring the survival of dense precolonial populations (White 1995; Woody 2003), yet commu-nity fishing practices also reveal instances of constraint imposed by this un-certainty, or the cyclical patterns of salmon runs. White (1995, 18) remarks that "the people who awaited the salmon were not simple fisher folk grate-fully taking the bounty of mother earth. Culturally, they made no assump-tions of the inevitability of the salmon's return. Their rituals, their social practices, their stories all recognized the possibility that the fish would fail to appear. They waited for salmon not with faith but with anxiety."

The arrival and harvest of salmon were firmly embedded in the social relations of precolonial indigenous societies. Bountiful seasonal salmon runs produced enough fish to ensure survival and perpetuate power and status for those who controlled access to the fishing grounds. Once sufficient numbers of salmon were obtained, the fishing ceased and salmon runs were allowed to pass upstream so that other communities could share in the bounty. This practice was also believed to confer good fortune on the community, ensuring the return of the salmon in future years.

Preservation of the salmon harvest was also structured along gender lines (White 1995). Women were prohibited from going near salmon harvest sites for fear that their presence might stop the salmon run. Given these restrictions, a gendered division of labor emerged whereby men procured the salmon and women assumed the vital task of preservation—drying, packing, and storing. These functional roles ensured collective survival. With the arrival of European colonialists, competition for the fish and commodification of salmon dramatically altered these practices. The subsequent reconstruction of salmon from a communal good to a commodity dramatically altered the fish and marked a watershed that continues to resonate in contemporary struggles.

Commercialization of Salmon Fisheries

Early colonial accounts of North America are replete with awestruck observations about the seemingly inexhaustible supply of fish (Bogue 2000; Taylor 1999; White 1995). Throughout North America this abundance of fish quickly entered into unfettered capitalist streams of development. Soon production was organized along industrial lines; industrialization of the fisheries thus transformed the food system as well as nature and society. In the hands of colonial agents, new tools and, later, science were employed effectively to perpetuate fish stock through hatcheries. As a result, colonialists transformed salmon into a commodity, thereby altering the precolonial structure of salmon. The extermination of indigenous "salmon people" underscored the structural transformation of salmon fisheries in North America (Taylor 1999, 134–36).

The commercialization of the Pacific salmon fisheries marked a key period in the social life of salmon. Under colonization, the extraction of natural resources intensified (e.g., gold, timber, transport), and salmon fisheries were no exception to these forces of capitalist accumulation. Early salmon fisheries in California deteriorated. In the search for new wealth, expansion moved to relatively untapped fisheries in the Pacific Northwest and Alaska. Technology became pivotal in this expansion. For example, in the development of commercial salmon fisheries, canneries adopted the use of more effective boats and fishing gear. Both entrepreneurs and hatcheries moved northward and, as Kelso elaborates, canneries began to drive the rise of commercial Alaskan salmon fisheries. On the Alaskan frontier, processors and canners were "the catalyst for the industry, enabling capital to create a new commodity by applying labor to vast runs of Pacific salmon" (Kelso 2000, 41). The unpredictability and geographic diffusion underlying these runs favored a highly mobile small-boat fishery. In the 1970s the state of Alaska initiated a limited entry program into the salmon fisheries that encouraged small, independent boat owners. Although the wild Alaskan salmon fishery is the world's largest, producing 20 to 30 percent of all salmon in the world, it is currently in a state of protracted economic decline brought on by competition with the farmed salmon industry. As a result, Alaskan fishers have become some of the fiercest critics of farmed salmon aquaculture.

Under the aegis of industrial capitalism, technology and science ushered in the structural transformation of salmon. Indigenous salmon fishers used tools and technology to ensure survival. Collective agency, as expressed in cultural relations—rituals, ceremonies, and taboos—were reflected in structured harvest practices and in the use of salmon platforms, weirs, traps, and processing in the postharvest phases. The technology deployed during the commercialization of salmon in the colonial era led to the rapid demise of the salmon fisheries. With salmon under increased fishing pressure and environmental assault, the establishment of hatcheries and canning technology transformed salmon from a subsistence crop to a commodity destined for external markets. In sum, one group of actors (indigenous fishers) was replaced by another set of actors (European commercial fishers). This structural transformation of salmon hinged on the application of technology and science. Significantly, scientific knowledge, as embedded in the advent of early hatchery methods, helped create the preconditions for another major form of structural transformation and intensified industrial production—aquaculture.

The Rise of Industrial Salmon Aquaculture: Feeding the World

Aquaculture is the controlled cultivation of aquatic organisms. Proponents of aquaculture contend that this method of production will bring about a whole host of social goods. These claims include aquaculture's ability to "feed the world," reduce seafood trade deficits, and create economic development opportunities, all the while conserving the ocean's wild fish stocks. Although aquaculture is an ancient form of animal husbandry, it was not until the mid-1980s that industrial farms began to emerge on a global scale (Bailey, Sinclair, and Jentoft 1996; Skladany 2000).

Salmon, one of a growing number of industrial aquaculture commodities, are well integrated into global production and consumption circuits of exchange. Farmed salmon have relatively high economic value and fit well with globalization's neoliberal projects. Temperate salmon aquaculture, like shrimp in the tropics, is often justified for its ability to create economic development in rural coastal communities and generate foreign exchange earnings for nation-states (Skladany 2000; Skladany and Harris 1995). As a result, a preliminary distinction between wild and farmed salmon sets the stage for what appears to be a protracted fight over seafood.

There is no definitive history of salmon aquaculture. It should be noted, however, that salmon hatchery production and scientific knowledge were well developed by the 1970s, when rural Norwegian residents began small-scale, open net-pen culture off the protected coasts of that country. Holm and Jentoft (1996, 27) note that in the 1970s the Norwegian government undertook rural development policies favorable to the settlement structure along the coast: "The Norwegian aquaculture industry [was] characterized by small, specialized firms geographically scattered in the coastal periphery. The industry was created by local pioneers who learned by doing. They took advantage of flexibility within social networks rather than control within hierarchies." Robust demand for fresh salmon positioned Norway as the undisputed leader in farmed production (see Table 7.1). By the 1980s, overproduction, failed government bailout schemes, and shifting policies favored a leaner shift to market mechanisms and the entry of large transnational corporations into salmon aquaculture. The rural cooperative aquaculture industry in Norway was absorbed into larger, vertically integrated corporations. These firms entered into joint ventures with foreign investors in Chile and Canada, among other countries, to capture an even larger market share (Kelso 2000; Weber 2003; Naylor, Eagle, and Smith 2003). Over the course of twenty years, farmed salmon became a global commodity under transnational corporate control.

Table 7.1 summarizes global salmon aquaculture production from 1985 to 2003. About 70 percent of all salmon in the world is farmed, with the remaining 30 percent being caught in the wild, mainly in Alaska.

In 2001, roughly thirty transnational corporations accounted for 70 percent of all farmed salmon. Nutreco, a subsidiary of British Petroleum, is the largest farmed salmon producer in the world (Weber 2003). Combined, the top six corporations—Nutreco, Netherlands; Pan Fish ASA, Norway; Fjord Seafood ASA, Norway; Stolt Sea Farm, Luxembourg; Cermaq ASA, Norway; and Aquachile S.A., Chile—control about 40 percent of the farmed salmon market (Naylor, Eagle, and Smith 2003; Weber 2003). Chile accelerated growth over the period 2001–4 and is soon expected to pass Norway as the world's leading producer. These conglomerates not only raise fish; they are also involved in the vertical integration of all components of salmon farming and other aquaculture species, such as tuna, cod, halibut, and trout (Weber 2003).

In the United States, virtually all farmed salmon are imported from Chile and Canada. Farmed salmon, mainly Atlantic, competes directly with wild salmon as smoked products, fresh steaks, and fresh fillets. In 2000, farmed Atlantic salmon brought an average wholesale price of $1.41 per pound, which stands in stark comparison to prices received by Alaskan salmon fishers (Weber 2003). For wild Chinook, Coho, and Sockeye, average 2001 wholesales prices were only $0.36–1.00 per pound (Meloy, Drouin, and Wyman 2002). The effect has been a protracted recession in the Alaskan salmon economy, with fishing permit values plummeting, fishing boats idling, and isolated fishing communities suffering economic hardship (Naylor, Eagle, and Smith 2003). Alaskan fishing groups, seafood marketing organizations, and the state government have undertaken economic bail-out programs and aggressive marketing and promotional efforts that have drawn attention to the unique features of wild-caught Alaskan salmon. The industry works hard to draw sharp distinctions between wild and farmed salmon, improve harvest quality, and decrease input costs associated with fishing in an effort to add value to their product (Kelso 2000; Naylor, Eagle, and Smith 2003). As of 2003,

Table 7.1   Estimated Global Salmon Aquaculture Production, 1985–2003 (in metric tons)*

Country	1985	1990	1995	2000	2003
Norway	30,000	146,000	281,000	437,000	507,000
Chile†	500	23,000	98,000	263,000	377,000
Scotland/Ireland	7,000	23,000	70,000	123,000	145,000
Canada	470	23,000	42,000	78,000	107,000
Faeroe Islands	540	12,000	8,000	28,000	56,000
United States	1,500	3,500	14,000	23,000	16,000
Other‡	8,500	39,000	41,000	51,000	33,000
Total	48,510	269,500	554,000	1,003,000	1,241,000

Source: Modified from Weber (2003, 9) and FAO (2003).

* Eighty-eight percent of global production consists predominantly of Atlantic salmon (*Salmo salar*), with some King or Chinook (*Onchorhynchus tshawytscha*) and Coho or Silver (*Onchorhynchus kisutch*). In general, Atlantic salmon are preferred by growers because of their docility and tolerance of high stocking densities (Weber 2003).

† Chilean production may be underestimated as of 2003.

‡ Includes Australia, New Zealand, Iceland, France, Spain, Greece, Turkey, Sweden, Finland and Portugal.

salmon stocks in Alaska, while robust and returning in record numbers, were still plagued by low prices. In many fishing communities, one would be hard pressed to find hope for a significant economic upturn in the foreseeable future.

Throughout the development of the global salmon aquaculture industry, both internal pressures and external problems have increased in magnitude. Internally, every industrial salmon aquaculture production region in the world has experienced periodic boom-and-bust cycles. Bankruptcies, consolidation, disease outbreaks, invasive escapes, contamination with antibiotics and toxic chemicals, legal battles, and significant sector restructuring in a glutted and volatile global market have proliferated. In this turbulent climate, the surviving transnationals have profited enormously. North American consumers find "fresh" salmon available year round, sometimes for less than $3 per pound. Externally, the whole system of open net-pen technology has been severely criticized by Alaskan fishers, Canadian environmental groups, coastal residents, First Nations, and, recently, some seafood consumer groups. These criticisms have concentrated largely on the environmental impacts of salmon farming, although health concerns, specifically the possible contamination of consumers by PCBs found in farmed fish, dealt a severe blow to the industry in 2004, as recorded by Hites et al. (2004). The industry responded by claiming that studies raising these concerns were "unscientific." They commenced a promotion of "organic" farmed salmon, which involves some minor changes in rearing methods. In the United States, some retailers adjusted quickly and began selling "organic" salmon in the hope that the label would assuage consumer fears.

Many of the environmental practices that are challenged pertain to the method of raising the fish. Salmon raised in open net-pens take advantage of the surrounding ocean. This practice passes environmental costs on to others outside the commodity chain. For example, a number of studies demonstrate how fish wastes pass through the nets. The heavy concentration of the excrement suffocates benthic life in the area surrounding the pen (Ellis and Associates 1996; Naylor, Eagle, and Smith 2003). Marine mammals that may eat farmed salmon can be shot or made deaf through the deployment of underwater acoustic devices near the pens (Morton 2002; Naylor, Eagle, and Smith 2003). Sea lice infestation from farms also poses a threat to wild stocks in the vicinity of the farmed salmon pens (Weber 2003; Naylor, Eagle, and Smith 2003). A final common criticism is that the colonization of British Columbia rivers by escaped Atlantic salmon threatens wild Pacific fish and ecological integrity (Volpe et al. 2000). These highly contentious debates are used by salmon farm advocates and critics alike to advance their respective positions.

Global Trajectories and "AquaAdvantages"

With aquaculture growing in global importance, both opposition to and defense of salmon aquaculture have mounted. An early environmental assessment of salmon aquaculture in British Columbia by Ellis and Associates (1996), funded by the Vancouver-based Suzuki Foundation, initiated a campaign that has exploded into a high-stakes conflict in Canada and in U.S. seafood markets. Over the past decade, two socially driven trajectories have emerged. First, scientific knowledge is viewed as the ultimate arbitrator for settling controversies. As a result, the proliferation of scientific reports, "expert" panel findings, and policy formulations have drawn considerable attention. More recently, corporate strategies and federal aquaculture development policy have begun to diversify from the salmon net-pen model.

Second, "counterexperts" in the form of environmental specialists, consumers, academics, and First Nations have challenged public-private funding allocation and research processes. They have identified gaps in the scientific knowledge and resulting policy. Initially these groups resisted the salmon industry and invested energy in developing marine conservation campaigns. This has allowed them to undertake a series of organized tactics to publicly influence social change (Khagram, Riker, and Sikkink 2002). As in other environmental arenas, they advocate a democratic approach to public understanding and participation in scientific matters (see Tesh 2000).

In short, an emerging multifaceted social arena has cohered where scientific momentum has sparked discourse over the extension of salmon aquaculture into new seafood production systems. Science-based "talks," debates, reports, and observations about the environmental and health effects of farmed salmon (e.g., endangered salmon, habitat deterioration and restoration, sea lice, Omega-3 proteins, antibiotics, toxic contamination) are the substantive discursive elements of the movement. These concerns then become the centerpieces of marketing campaigns in an effort to distinguish between wild and farmed fish for the reflexive consumer. The further development of new aquaculture offshoots instigated by science, corporate strategies, and aggressive aquaculture policy formulations has greatly expanded this arena. As a result, these global trajectories become the essence of corporate strategy and are vigorously countered by movement resistance. They include:

- Increasing the size, scale, and geographical location of salmon farms.
- Diversifying into new species such as bluefin tuna, cod, halibut, and cobia, using variations of the salmon net-pen model.
- Propositions to develop U.S. fish farms in the 3–200 mile U.S. Exclusive Economic Zone.
- Calculating the use of genetically engineered AquaAdvantage salmon.

Kelso (2000) predicts that salmon farm size may have to increase significantly in order to capture efficient economies of scale. To an extent, this has already begun, as transnational corporations have consolidated and expanded their holdings, diversified locations, and globally speculated on production sites that hold the greatest market advantage (Raincoast Conservation Society 2004). With salmon net-pen technology established and overproduction a chronic problem, growers are diversifying into other carnivorous species. As of 2004, there are approximately forty sablefish (black cod) and seven halibut farms currently operating in British Columbia (Intrafish 2004). In other Pacific Ocean areas, bluefin tuna "ranching" involves catching wild tuna, holding them in net-pens for "fattening," and then sending them to the high-end and seemingly insatiable sushi markets in Japan and the United States (Weber 2003).

Open-ocean aquaculture—the growing of finfish under controlled conditions in high-tech submerged cages within the 3–200 mile U.S. Exclusive Economic Zone—is under way, with legislation proposed to streamline permitting and private property rights for full commercial application (Skladany, Belton, and Clausen 2005). The U.S. Department of Commerce, through the National Oceanic and Atmospheric Administration (NOAA) and Sea Grant Program, is developing legislation and technology for siting fish farms away from coastal areas in exposed ocean conditions.

Finally, the pending release of the "AquaAdvantage" genetically engineered (GE) salmon, designed by the U.S-Canadian firm A/F Protein, Inc., carries untold political, social, and environmental consequences. Dubbed "frankenfish" by critics, this GE fish is the object of derision and concerted resistance by a number of environmental, consumer, and citizen watchdog groups (see the previous chapter by Munro and Schurman). Beginning in the mid-1980s, the transfer of select foreign genes (i.e., growth genes, antifreeze genes) into the host genome of commercial aquaculture species such as Atlantic salmon has been refined for speedier growth and freeze tolerance (Skladany 2000). In the previous chapter, Munro and Schurman conclude that "anti-GE activists have catalyzed important new regulatory restraints on the technology" and have helped turn many consumers away from genetically engineered food. To an extent, this has happened with the AquaAdvantage GE salmon, at least in principle. Although more than fifty GE fish have entered the world food system, salmon draw an inordinate amount of attention because of their prominence in cultural and economic affairs (Skladany 2000).

Initially, the U.S. government was poorly prepared to respond to earlier GE fish research developments. A proprietary application for release of the AquaAdvantage salmon was put under the purview of the Food and Drug Administration (Skladany 2000). Over the past two years, the AquaAdvantage salmon has been evaluated as a "new animal drug" in terms of being safe for human consumption. Critics argue that the FDA lacks environmental assessment expertise in this area. In one hypothetical scenario, the escape of GE salmon would threaten endangered wild Atlantic salmon stocks through a "Trojan gene effect"—whereby entry of the GE fish into a wild fish population would lead to their extirpation (see Muir and Howard 1999). Moreover, Kelso (2000) argues that the doubled growth rate displayed by AquaAdvantage salmon would transform the industrial structure of salmon farming owing to competitive advantages that would benefit those in control of the GE fish. In short, the GE fish would lower variable input costs (e.g., feed). At this writing, it remains to be seen how the FDA will rule upon these matters. Overall, these corporate diversification strategies have inspired a far-flung and eclectic movement opposed to industrial aquaculture. Let us turn now to some of these emerging salmon and seafood movements and consider the transformation of the seafood system in terms of human agency.

The Movement Against Salmon Farming: Resistance as Agency

The case of organized protest in the aquaculture arena reminds us that resistance is one form of agency, but it comes in many varieties. To speak of "the seafood movement" is to gloss over the diversity of movement repertoires and ideologies that make up a sundry array of actors. To comprehend their range of responses we must first acknowledge that wild and farmed salmon stand in stark contrast to some land-based agrifood production systems. Salmon draw an inordinate amount of attention from diverse social and cultural standpoints. Salmon are multifaceted cultural entities owing to their interactions with different actors in our food system. As a result, human agency is bound up with diverse interpretations of where salmon should be positioned in our food system. For some movement actors, "wild" salmon are environmentally charged symbols, highly romanticized, often evoking an identification beyond that of a mere food item. As a result, human agents express partial, eclectic, and at times overlapping efforts to save "wild" salmon and oppose farmed salmon and related offshoots such as open-ocean aquaculture and genetically engineered salmon. Human agency is also expressed by environmental groups in their efforts to inform consumers about sustainable seafood buying practices, for example, by promoting "wild" salmon or species that are "in season." The aquaculture industry also exerts agency when they endeavor to inform consumers that their product is healthy, safe, and environmentally friendly. Fundamentally, movement campaigns struggle for access to, and construction of, wild salmon and, by means of extension, nature.

More concretely, wild salmon are vital for cultural survival in Pacific Northwest fishing communities. In contrast, farmed salmon are industrially produced food commodities that in British Columbia threaten cultural survival. The salmon aquaculture industry *assumes* that farming takes fishing pressure off wild salmon, thereby preserving them, while bringing economic development to isolated coastal communities. These claims, however, have not been well substantiated—a point that critics are quick to make. All of these social forces suggest uncertain futures for salmon and for those whose livelihoods and cultures are wedded to the fish.

We can detect a number of emerging movement themes that offer insight into the possible future of salmon. These include:

Photo: Doug Sipes

- *Environmental and Health Concerns.* Charges that farmed salmon pollute the water, threaten wild salmon, and are unsafe to eat have been made and countered by an industry that claims otherwise. The salmon aquaculture industry took a major blow when a scientific study (Hites et al. 2004) revealed that farmed salmon contain higher levels of PCBs and dioxins than wild fish.
- *Sustainable Livelihoods and Cultural Survival.* Salmon fishers, including First Nations in the Pacific Northwest, have been hurt economically and culturally by the explosion of fish farms. Industry counters that jobs and economic development benefits have been created.
- *Markets as an Arena for Influencing Social Change.* A number of marine conservation foundations have funded "market campaigns" that seek to influence consumer choices about salmon and other seafood. Some campaigns address "sustainable seafood choices" over a range of marine species, while others advocate "wild" salmon as a direct substitute for farmed fish. Chefs, nutritionists, and other health professionals have also begun to weigh in on consumption debates (Vandergeest and Skladany 2002).

A number of North American environmental and affiliated organizations actively campaign against farmed salmon. Opposing these groups is a well-funded salmon aquaculture industry backed by government policies that support industrial aquaculture and bring science to bear on a wide range of contentious issues. The most entrenched resistance against the current net-pen rearing of farmed salmon can be found in British Columbia. I discuss the British Columbian–based Coastal Alliance for Aquaculture Reform (CAAR) below. Although the CAAR coalition is focused on salmon campaigning in British Columbia, I also examine some of the more broadly focused "seafood-choice" campaigns, because this emerging seafood movement has direct influence on specific campaigns such as CAAR. I conclude the discussion by examining the potential for transforming the seafood system and the exercise of agency.

The CAAR Campaign

The British Columbian Coastal Alliance for Aquaculture Reform is a coalition of seven Canadian environmental groups and three First Nations. With the lifting of a moratorium on new farms by the British Colombia

(BC) provincial government in October 2002, CAAR swiftly mounted a highly focused campaign to reform salmon farming practices. In sum, CAAR states that farmed salmon will be safe when the fish-farming industry

- uses technology that eliminates the risks of disease transfer and fish escapes
- guarantees that waste is not released into the ocean
- labels its fish "farmed" so consumers can make informed choices
- develops fish feed that doesn't deplete global fish stocks
- ensures that wildlife is not harmed as a result of fish farming
- prohibits the use of genetically modified fish
- eliminates the use of antibiotics in fish farming
- ensures that contaminants in farmed fish don't exceed safe levels
- stops locating fish farms in areas opposed by aboriginal groups or other local communities (CAAR 2004)

The solution, according to CAAR, involves developing closed containment facilities on land. With most BC salmon exported to the United States, CAAR is also mounting market campaigns to target consumers in select West Coast cities and at certain retail outlets. The CAAR coalition encourages consumers to avoid farmed salmon and make sustainable seafood choices, although not exclusively substituting wild for farmed salmon because of the endangered status of many wild stocks. The conduit for raising public awareness in Canada rests on the almost relentless monitoring of on-site industry developments, negotiating with industry, and challenging official science that is seen as pro-industry. CAAR has also become a fierce critic of the selective deployment of information in the United States, using print and electronic media to send mass messages, in the hope of discouraging consumers from buying farmed BC salmon. Communication is tailored to fit specific cultural geographies. For example, in British Columbia, the challenge to government-industry science, research, and policy is featured prominently in the *Vancouver Sun* and smaller dailies throughout the province. Controversies such as the infestation of wild salmon by sea lice make front-page news. In the U.S. Midwest, activists draw an analogy between the "floating salmon feedlots" and confined animal feeding operations (CAFOs) in an attempt to connect the salmon issue to agri-food conflicts specific to other areas.

Journalists play a vital role in publicizing new science-based issues and controversies, and in effect are the new knowledge brokers, thanks to their almost instantaneous transmission of news and feature stories. For example, an in-depth exposé may take no more than one month to complete. Online electronic seafood news services such as the Wave and Intrafish often are the first to break stories on salmon fisheries and aquaculture issues. Such campaigns against salmon farming have had an impact on sales of farmed salmon, U.S. wild salmon prices, and the entry of salmon-farming issues into the political arena.

Under this steady barrage of criticism and eroding consumer preferences for farmed salmon, the British Columbia Salmon Farmers Association hired the international media giant Hill and Knowlton in 2003 to counter criticism and spin the public image of the industry in a more favorable light. Hence, one sees a steady flow of press releases, electronic newsletters, editorials, exposés, investigative reports, and letter-writing campaigns in BC and U.S. media outlets. Other industry organizations, such as Salmon of the Americas and the Positive Awareness Society for Aquaculture, champion the benefits of salmon aquaculture in terms of health, economic development, and job creation—even holding counterrallies at BC fish farm sites. Even so, this is not a simple matter of pitting "environmentalists" against the salmon industry and its backers in the BC provincial government. Weighing in on this increasingly contested issue are small-scale fishers, recreational angling groups, academics, the tourism industry, direct action networks, old growth forest activists, artists, naturalists, and a well-educated citizenry.

The position of First Nations on salmon aquaculture is especially noteworthy. As Kelso (2000) observes, First Nations are almost unanimously opposed to salmon aquaculture in British Columbia. Given the ongoing struggle for sovereignty, First Nations may play the decisive role in the whole social ordering of salmon owing to their legal capacity to make significant structural changes. For instance, First Nations have the power to enforce marine treaty rights and thereby structure human use of the environment and access to wild salmon stocks. There is some disagreement among First Nations, however, with a few British Columbian tribes actively involved in net-pen salmon aquaculture (Kelso 2000; Naylor, Eagle, and Smith 2003).

By contrast, nonaboriginal Alaskan fishers are totally enmeshed in global market mechanisms beyond their control. Their exercise of agency confronts enormous structural constraints. As a result, they struggle to compete against a tide that threatens to eliminate their livelihoods and ways of life. It remains to be seen whether consumer education campaigns will continue to receive critical foundation support for driving reform in the BC salmon farm industry and elsewhere. CAAR, the National Environmental Trust (NET), and Seaweb recently committed significant funds to a U.S.-based salmon market campaign. This effort promises to be transnational in scope and will work primarily at the stockholder and distributor level to urge the adoption of rigorous environmental standards concerning farmed salmon.

Seafood Consumer Education Campaigns

The CAAR campaign and the recently funded NET-Seaweb salmon markets campaign are part of a much broader sustainable seafood consumption movement. Similar efforts are under way in the United States (Ecotrust, Save Our Salmon), Chile (Oceanas, Terran Foundation), Scotland (Salmon Farm Protest Group), and along the Maine-Canadian coastline (East Penobscot Bay Environmental Alliance). Not all of these groups take the wild-versus-farmed debate as a point of departure. For example, Chile has no wild salmon. The efforts there are focused primarily on sustainable fishing livelihoods and labor conditions in the transnational salmon-processing facilities. The CAAR campaign does not explicitly endorse wild over farmed salmon when it comes to seafood choices, yet Alaskan fishing groups do. Generally, fishing groups are known to take a hard line against salmon farms and other newly emerging forms of fish farming, such as black cod and open-ocean aquaculture. From time to time these groups align with environmental organizations when their interests are served. At the same time, large marine conservation foundations typically are highly critical of almost any form of commercial fishing, as are some of the major environmental groups. In the middle of the road are well-funded groups like the World Wildlife Fund (WWF), which are engaged in "salmon aquaculture dialogues" with the global salmon industry. These dialogues have created turbulence among environmental actors opposed to farmed salmon, who see the WWF as being co-opted by industry.

The emerging seafood movement is not confined to salmon but also includes U.S.- and Canadian-based marine conservation and culinary organizations (e.g., the WWF, Monterey Bay Aquarium, Chefs Collaborative, Seafood Choices Alliance, and the Endangered Fish Alliance). These organizations advocate making "sustainable" seafood consumption choices on the basis of environmental assessments. The majority of campaign efforts are directed toward upper-middle-class consumers and designed to raise awareness in public venues and at high-end restaurants. Chefs are increasingly viewed as the key link between the environmental conservation movement and the high-end consumers who frequent these establishments. As Johnston

(Chapter 4 in this volume) notes, celebrity chefs have become highly influential in food circles, and they have the same status in the debate over wild versus farmed salmon and other aquaculture practices. The *Vancouver Sun* reported on a highly publicized trip by two celebrity chefs to British Columbia:

> Kennedy and Moonen are the culinary equivalent of rock stars. When they speak, the restaurant world listens. And their words are heating up the salmon-farming spin wars. Kennedy hasn't served farmed salmon in five years, but the trip west was long overdue. . . . "There are more and more sad tales associated with fish farming, and from a taste point of view, it's not the same." Kennedy, who operates JK Wine Bar in Toronto, is a trail-blazing chef with huge influence and trendsetting clout. Moonen is chef/owner of one of New York's hottest seafood spots, Restaurant RM. Both Moonen and Kennedy are boycotting farmed salmon until the highest standards for aquaculture are met. (Stainsby 2004)

Marine conservation and culinary groups regularly host public seafood taste tests and discussion panels. They also produce technical reports and handy wallet-sized seafood shopping cards for the reflexive consumer interested in making informed sustainable choices. Ocean conservation and seafood consumption also extend to sustainable marine fisheries certification through the Marine Stewardship Council (MSC). So far, the MSC has certified a number of marine fisheries as sustainable; the Alaskan salmon fisheries were the first, in 1997. In sum, a number of groups work on salmon and related marine conservation issues under a very broad umbrella. At times these groups can oppose one another, as is the case with the World Wildlife Fund, but it is very clear that the sustainable seafood movement seeks to penetrate mainstream markets. Above all, foundation support is absolutely vital to sustaining these campaigns and to "disciplining" and aligning the movement into a coherent whole.

The Context of Human Agency in the Case of Salmon

Salmon and other fish species are on the cusp of an emerging seafood movement and, by extension, ocean conservation movement. This movement comprises an eclectic group of actors all vying to influence the structure of seafood production and consumption. Some of them see the preservation of wild salmon as an essential ecological and democratic imperative consistent with a highly romanticized view of nature. Other advocates of wild salmon, such as commercial fishers, are most concerned with defending their way of life against market inroads by salmon aquaculture. We are once again at a crossroads in the transformation of salmon, and the structure of seafood production and consumption, through collective action and reflexive consumption patterns. The future of salmon and sustainable seafood choices remains a matter of great contention and growing visibility in the public arena.

The debate over the future of salmon in particular, or of seafood more generally, illustrates the desire to use, interpret, and advocate for more *democratic* science-based solutions to pressing contemporary issues. In the CAAR campaign, one set of demands centers on developing science and technology for closed-containment rearing of salmon. At the same time, CAAR has consistently challenged official Department of Fisheries and Oceans and industry-based science over issues such as sea lice infestation of wild salmon from fish farms, antibiotic use, contaminants, and the escape of farmed Atlantic salmon into British Columbian waters. While the outcomes of scientific research are uncertain, CAAR has made the strategic decision to mount marketing campaigns against BC farmed salmon in the United States. Other groups make use of fisheries management findings to advocate for science-based ocean conservation of fish stocks. These groups, quite literally, usher chefs to the table and to upscale retail outlets to educate consumers on how to make sustainable seafood choices.

In a remarkably concise book on environmental activism and science, Tesh (2000) argues that "the environmental movement, like all social movements, is far larger than the organizations it has spawned. . . . Unless one includes in the term *social movement* all the movement's actors—the whole gamut of men and women who invent and champion the movement's principles—one is likely to miss a movement's most important contribution . . . these include the writers and scientists . . . as well as a host of journalists, politicians, bureaucrats, educators, artists and others who use their professional skills to foster the movement" (120–21). Such is the case with salmon. Structural change over salmon entails a diverse array of movement actors and a highly malleable arena for expressing human agency. These agents bring experiential knowledge and abundant resources (e.g., organizational acumen and foundation support) to this arena. They couple their action with an astute reading of scientific literature in order to articulate their positions on salmon. For example, Canadian First Nations' almost total resistance to salmon aquaculture rests on the immediate threat to wild salmon stocks. The rationale for resisting farmed salmon is based on fisher observations, cultural traditions, scientific findings, traditional livelihoods, and, more fundamentally, national sovereignty. We can surmise that, in the case of salmon, human agency is actively shaping the structure of production and consumption, even while it is subject to preexisting constraints imposed by that structure.

In the case of salmon, the coextensive interplay between structure and agency has resulted in a reduced sense of culture and diminished livelihoods for some and a new form of wealth for others. When movement actors, including corporate aquaculturists, champion wild or farmed salmon, they also implicitly articulate a view of nature and society. Actors define themselves in relation to salmon—as a miraculous gift from nature, as the heart and soul of coastal economies, as a spiritual entity in a cosmological order, or as a commodity grown in crowded net-pens destined for global market transactions.

I would cautiously suggest that, in the case of salmon, organized citizen involvement would improve our collective understanding of the seafood system. As a case in point, in the autumn of 2004 I participated in a "Salmon Nation" block party hosted by the Portland-based Ecotrust organization. Held at Ecotrust's recently renovated "green building" in the downtown area, the block party brought together a diverse array of people who celebrated wild salmon. Local artisans sold a variety of crafts at individual booths, and locally grown food—including the prized Columbia River salmon obtained from the Columbia River Intertribal Fisheries Commission—was featured on the menu. Live music occupied center stage, and performance artists mingled among the thousands of people in attendance. In this venue, everyday citizens pledged to engage in reflexive consumption and to become members of "Salmon Nation."

Featured talks and discussion emphasized the value of salmon to the region and underscored close community ties with salmon—as food, as spirit, as life. The connection between salmon and a healthy food system was manifested in many different ways by citizens from all walks of life. At stake here was an almost spiritual tie between salmon and group identity, seen in the way participants situated themselves in relation to salmon. Some may charge that an overly romanticized view of nature and social relations animated the event. It can also be said that such gatherings, and the spirit they generate, may hold the key to a better understanding of salmon, society, and human agency.

This potential was best stated by Ed Hunt of Ecotrust in his response to the question "what is Salmon Nation?" Salmon Nation, Hunt explained, "is a geographic landscape as well as a landscape of mind. It is a nation that does not yet exist. It is a nation that has always been. It is a cultural identity built around the soil and streams touched by Pacific salmon. It stretches from Alaska down to California and inland to Idaho and even Montana. . . . It is a landscape that goes by different names these days—the Pacific Northwest, Cascadia, the Rain Forest Coast. Yet salmon were here before people ever were, and the first people knew the districts of this nation long before the mountains were named for English nobility. Those people were rich because of the salmon. Indeed the very lands of Salmon Nation have been enriched by the bodies of salmon over the millennia. Beyond salmon, we are bound in this region by other issues, by water, by power, by trade and history. We face common problems, share common interests, and look to each other with a common history."[2]

Perhaps a "Salmon Nation" can begin to broaden and even enlighten thinking about salmon and humans' duty to nature and society through our respective seafood choices. A "Salmon Nation" represents a provocative first step in changing the current seafood system.

Conclusion

This case study demonstrates that the approach to, use of, and understanding of the ocean as a source of food has shifted significantly over the past few decades. From earlier colonial perceptions about "infinite" oceanic abundance coupled with growing technological development in harvesting seafood, a series of transformations occurred, beginning with the commercialization of fishing in North America. Technology and science mediated these expressions of human agency and transformed the structure of seafood production and consumption. More recently, a key development is the emergence of salmon aquaculture. In the case of farmed salmon and the broader seafood consumption movement, a number of actors continue to work to transform current salmon production and consumption networks.

One consequence of the movement for sustainable seafood is the reconstruction of salmon and actor identity in ways that consequentially redefine relations with nature, seafood, and society. In this respect, the sustainable seafood movement is attempting to introduce new environmental criteria concerning sourcing and consumer choice. Will it succeed?

2. See www.ecotrust.org/.

The case of salmon is important because it illuminates a number of broader emerging scenarios. First, environmental groups and indigenous groups have made inroads with consumers concerning farming practices that they view as unsustainable. With environmental groups, the effort is highly dependant on the major marine conservation foundations for the long-term funding support that drives consumer education campaigns. Second, the sustainable seafood movement faces a vast multibillion dollar seafood-aquaculture industry that has become more reflexive and cost conscious about environmental and social problems associated with unsustainable sourcing and human health issues. It is also very clear that aquaculture will continue to play a growing role in seafood production. Finally, a convergence of the environmental movement and the seafood industry is taking place around the adoption of environmental standards, certification, and sustainable production. It remains to be seen, however, whether policy goals can be realized in practice.

This case study of salmon and sustainable seafood illustrates an expanding arena for the exercise of human agency. With the commercialization of salmon fisheries and the recent rise of global salmon aquaculture, human agents have transformed the structure of salmon production and consumption. Although there are pronounced structural constraints to the exercise of human agency, there are numerous arenas where the conscious shaping of salmon and identity are occurring in ways that are eclectic, contentious, and due for a protracted social struggle. This restructuring is not unique to salmon. Other oceanic seafood production likewise faces market conditions unlike those of land-based agrifood systems. The distinction between wild and farmed salmon offers a hint as to where agency will be expressed in broader structural arenas. But the outcome of an alternative sustainable seafood system remains highly uncertain.

REFERENCES

Bailey, Conner, Peter Sinclair, and Svein Jentoft, eds. 1996. *Aquaculture Development: Social Dimensions of an Emerging Industry.* Boulder, Colo.: Westview Press.

Bogue, Margaret Beattie. 2000. *Fishing the Great Lakes: An Environmental History, 1783–1933.* Madison: University of Wisconsin Press.

Callon, Michel, and John Law. 1997. "After the Individual in Society: Lessons on Collectivity from Science, Technology, and Society." *Canadian Journal of Sociology* 22: 165–82.

Coastal Alliance for Aquaculture Reform (CAAR). 2004. "Think Twice About Eating Farmed Salmon." Farmed and Dangerous Campaign Brochure. www.farmedand dangerous.org/.

Ellis and Associates. 1996. "Net Loss: The Salmon Netcage Industry in British Columbia." Vancouver: David Suzuki Foundation. www.davidsuzuki.org/.

Food and Agriculture Organization of the United Nations (FAO). 2003. *Fishery Statistics Yearbook, Aquaculture Production 2003.* www.fao.org/.

Hites, Ronald A., Jeffrey A. Foran, David O. Carpenter, M. Coreen Hamilton, Barbara A. Knuth, and Steven J. Schwager. 2004. "Global Assessment of Organic Contaminants in Farmed Salmon." *Science* 303 (January): 226–29.

Holm, Peter, and Svein Jentoft. 1996. "The Sky Is the Limit? The Rise and Fall of Norwegian Salmon Aquaculture, 1970–1990." In *Aquacultural Development: Social Dimensions of an Emerging Industry,* ed. Conner Bailey, Svein Jentoft, and Peter Sinclair, 23–42. Boulder, Colo.: Westview Press.

Hunt, Ed. 2003. "What Is Salmon Nation?" www.ecotrust.org/.

Intrafish. 2004. "U.S. Senator Asks Canada to Delay Black Cod Farming." Intrafish Media Network, 17 August 2004. www.intrafish.org/.

Kelso, Dennis Doyle Takahashi. 2000. "Aquarian Transitions: Technological Change, Environmental Uncertainty, and Salmon Production on North America's Pacific Coast." PhD diss., University of California, Berkeley, Department of Energy and Resources.

Khagram, Sanjeev, James V. Riker, and Kathryn Sikkink. 2002. "From Seattle to Santiago: Transnational Advocacy Groups Restructuing World Politics." In *Restructuring World Politics: Transnational Social Movements, Networks, and Norms,* ed. Sanjeev Khagram, James V. Riker, and Kathryn Sikkink, 3–23. Minneapolis: University of Minnesota Press.

Manning, Richard. 2003. "Ghost Town." In *Salmon Nation: People, Fish, and Our Common Heritage,* ed. Edward C. Wolf and Seth Zuckerman, 33–44. Portland, Ore.: Ecotrust.

Meloy, Buck, Michel Drouin, and Jeb Wyman. 2002. "Salmon: 2002 Stats Pack—The State of the Industry." *Pacific Fishing Yearbook* 23 (3): 22–33.

Morton, Alexandra. 2002. *Listening to Whales: What the Orcas Have Taught Us.* New York: Ballantine Books.

Muir, William M., and Richard D. Howard. 1999. "Possible Ecological Risks of Transgenic Organism Release When Transgenes Affect Mating Success: Sexual Selection and the Trojan Gene Hypothesis." *Proceedings of the National Academy of Sciences USA* 96 (November): 13853–56.

Naylor, Rosamond L., Josh Eagle, and Whitney L. Smith. 2003. "Salmon Aquaculture in the Pacific Northwest: A Global Industry." *Environment* 45 (8): 18–39.

Raincoast Conservation Society. 2004. "Salmon Stakes: An Investigative Report on the Multinational Companies That Control B.C.'s Salmon Farming Industry." Manuscript.

Skladany, Mike. 2000. "Sociology, Science, and Technology: The Case of Transgenic Fish Research and Development." PhD diss., Michigan State University, Department of Sociology.

Skladany, Mike, Ben Belton, and Rebecca Clausen. 2005. "Out of Sight and Out of Mind: A New Oceanic Imperialism." *Monthly Review* 56 (9): 14–24.

Skladany, Mike, and Craig K. Harris. 1995. "On Global Pond: International Development and Commodity Chains in the Shrimp Industry." In *Food and Agrarian Orders in the World-Economy,* ed. Philip McMichael, 169–91. Westport, Conn.: Praeger.

Stainsby, Mia. 2004. "Culinary Standouts Are Determined to Change the Way the World Views Farmed Fish." *Vancouver Sun,* 7 July.

Taylor, Joseph E., III. 1999. *Making Salmon: An Environmental History of the Northwest Fisheries Crisis*. Seattle: University of Washington Press.

Tesh, Sylvia Noble. 2000. *Uncertain Hazards: Environmental Activists and Scientific Proof*. Ithaca: Cornell University Press.

Vandergeest, Peter, and Mike Skladany. 2002. "Consumer-Driven Aquaculture: How Food Politics Might Impact Industrial Shrimp and Salmon Aquaculture." Paper presented at the sixty-fifth annual meeting of the Rural Sociological Society, Chicago, 15 August.

Volpe, J. P., E. B. Taylor, D. W. Rimmer, and B. W. Glickman. 2000. "Evidence of Natural Reproduction of Aquaculture-Escaped Atlantic Salmon in a British Columbia River." *Conservation Biology* 14 (3): 899–903.

Weber, Michael L. 2003. "What Price Farmed Fish: A Review of the Environmental and Social Costs of Farming Carnivorous Fish." Report prepared for the SeaWeb Aquaculture Clearinghouse, Providence, R.I. www.seaweb.org/resources/reports.php #crossroads.

White, Richard. 1995. *The Organic Machine: The Remaking of the Columbia River*. New York: Hill and Wang.

Woody, Elizabeth. 2003. "Recalling Celilo." In *Salmon Nation: People, Fish, and Our Common Heritage*, ed. Edward C. Wolf and Seth Zuckerman, 9–16. Portland, Ore.: Ecotrust.

INSTRUCTOR'S RESOURCES

Key Concepts and Terms:
1. First Nations
2. Aquaculture
3. Social Movement
4. Sustainable seafood choices
5. Salmon Nation

Discussion Questions:
1. How have salmon been transformed? What role have tools, technology, and science played in transforming salmon?
2. What is Salmon Nation? How does Salmon Nation spark human agency and identity redefinition for salmon and for people?
3. Design a salmon advocacy campaign (either "wild" or "farmed"), making use of ideas and elements from this chapter.

Salmon Aquaculture Videos:
1. *Net Loss: The Storm over Salmon Farming*. Bullfrog Films, 2003 (52 minutes).
2. *Farming the Seas*. Habitat Media, 2004 (60 minutes).

Salmon Aquaculture on the Internet:
1. Coastal Alliance for Aquaculture Reform: http://farmedanddangerous.org/.
2. Seafood Choices Alliance: http://seafoodchoices.org/.
3. Ecotrust: www.ecotrust.org/.
4. British Columbia Salmon Farmers Association: www.salmonfarmers.org/.
5. Salmon of the Americas: www.salmonoftheamericas.com/.
6. Pure Salmon Campaign: http://puresalmon.org.

Additional Readings:

1. Barcott, Bruce. 2001. "Aquaculture's Troubled Harvest." *Mother Jones,* November–December. www.motherjones.com/news/feature/2001/11/aquaculture.html.

2. Ecotrust. "The Hidden Costs of Farmed Salmon" and "Salmon Nation." www.sectionz.info/.

3. Skladany, Mike, and Peter Vandergeest. 2004. "On the Menu." *Alternatives Journal* 30 (2): 24–26.

4. Wolf, Edward C., and Seth Zuckerman. 2002. *Salmon Nation: People, Fish, and Our Common Heritage.* Portland: Ecotrust.

PART III

CONSTRAINTS TO AGENCY

8

INFERTILE GROUND: THE STRUGGLE FOR A NEW PUERTO RICAN FOOD SYSTEM

Amy Guptill

Since the late 1970s, activists in Puerto Rico have struggled to promote sustainable agriculture as a means of breaking the island's dependence on both food and food stamps from the United States. Farmers and activists have established farmers' markets and other direct-marketing opportunities, organized workshops on organic techniques and agroecology, and created a channel through which to lobby for government support for organic agriculture development. The movement for a more sustainable and local agriculture grew out of environmentalism, which was galvanized in the 1960s during a successful campaign to prevent a strip-mining operation on the island (Concepción 1995). That experience led many activists to emphasize that the struggle for environmental integrity is part of a broader struggle against the colonial relationship of Puerto Rico to the United States (Concepción 1995; Berman-Santana 2000).

The growth of the global organic sector in recent decades has provided sustainable agriculture activists in Puerto Rico with a new conceptual and economic framework for combining their ecological and political concerns. A recent report for the International Federation of Organic Agriculture Movements found that "organic agriculture is practiced in almost all countries of the world, and its shares of agricultural land and farms is growing

An earlier version of this paper was presented at the conference "Agrarian Struggles and Agro-Food Re-Regulation," Minneapolis, 8 June 2001. I would like to thank Maria Benedetti, Vivian Carro-Figueroa, Gary Green, Phil McMichael, Kai Schafft, Betty Quick, Wynne Wright, Gerad Middendorf, and several anonymous reviewers for helpful comments. The late Tom Lyson was an invaluable mentor throughout this project. Support for this research was provided by the Einaudi Center for International Studies, Latin American Studies Program, International Studies in Planning, Cornell Institute for International Food and Agriculture Development, and the Andrew W. Mellon Student Research Award, all at Cornell University.

everywhere" (Yussefi and Willer 2002, 9). In the United States alone, consumption of organic and other "natural" products nearly tripled between 1993 and 1998 from $6 billion to $17 billion (Wellman 2000). The Dominican Republic, Costa Rica, and Guatemala are leading regional organic producers of both traditional tropical commodities like coffee, bananas, cacao, sugar, and nontraditional commodities like counterseasonal vegetables (Raynolds 2000; Yussefi and Willer 2002). Organic products are also a small but growing part of the agricultural sectors of Belize, Cuba, El Salvador, Honduras, Nicaragua, Trinidad and Tobago, and Suriname. Among Caribbean islands, Puerto Rico is perhaps best positioned for organics, because agricultural production declined rapidly after World War II, leaving extensive agricultural acreage that has never been treated with agrochemicals, while a modern transportation and energy infrastructure was created.

Through commitment, hard work, and careful strategizing, activists have achieved a few notable successes, but, surprisingly, organic farming is still negligible at best. In 2000–2001 there were at most twelve alternative growers who depended on agricultural sales for at least 25 percent of their income. Another twenty or so retirees and other hobbyists owned farms and expressed interest in organics, but did not depend on farming for their livelihood. Comparatively, there were six organizations whose primary missions were to promote organic or sustainable agriculture and six others that pursued similar efforts as secondary goals. Thus, there were as many organizations promoting sustainable agriculture as there were farmers. Similarly, the movement was large (with about 150 consistent participants, and many more occasional ones) compared to the small number of farmers. Overall, then, while activists have made important gains in connecting producers, consumers, academics, and some policymakers around the promotion of organics and related alternatives, the movement has not succeeded in establishing a vigorous and growing farming sector.

The purpose of this chapter is to examine how the structure of the Puerto Rican food system both enables and constrains the agency of activists seeking to transform the agrifood landscape. I first provide a brief historical narrative of the structure of the Puerto Rican food system to explain how Puerto Rico came to have a food system that is almost entirely dependent on both food imports and nutritional assistance programs. I turn next to an account of recent prominent efforts to effect food system change, all of which encountered major obstacles to transformative agency. The structure of the agrifood system means that contemporary activists face a bigger task than simply diverting agrifood resources to sustainable sectors. Rather, their task is to entirely regenerate a productive working landscape and a locally grounded food system (see Guptill 2004).

The experiences of contemporary activists demonstrate that this task is fraught with formidable challenges. The following section of the chapter describes four key events in recent sustainable agriculture efforts and interprets the obstacles they encountered as products of a dependent food system structure that marginalizes production. The key events are based on data gathered in fifteen months of ethnographic field research between 1998 and 2001, as well as a two-week follow-up visit in 2005. While the particularities of each case are open to multiple interpretations, an approach of "solving for pattern" (Lappe and Lappe 2002) suggests that the diverse obstacles reflect a shared context.

From that pattern-based perspective, the research findings presented here are consistent with two general points about agency in the agrifood system. First, like other chapters in this volume (particularly the following chapter by Jussaume and Kondoh), these findings highlight the ways in which the place-specific context both enables and constrains agency for activists seeking to effect lasting, broad-scale change. Second, on a conceptual level, these findings support the notion that a workable theory of agency must explain the relationships among the desire for change, strategies for change, and ultimate success. In Friedland's language (Chapter 2 in this volume), some individuals and groups who are consciously (and socially) dissatisfied with the status quo and act to create alternative networks may still fail to create change in a tangible, proximate timeframe. In other words, the findings presented here raise a question: do efforts count as "agency" only when they succeed in direct, easily recognizable ways? The struggles of sustainable agriculture activists in the face of enormously challenging conditions in Puerto Rico suggest that an inclusive, multidimensional understanding of agency is needed if we are to appreciate the significance of their work.

Origins of a Dependent Food System

Like other islands in the Caribbean, Puerto Rico has long been a place that produces what it does not consume and consumes what it does not produce (Mintz 1985). This pattern, begun under the Spanish, was re-created during the early years of U.S. rule (1898–1930) and again through the postwar industrialization program. In the four hundred years of Spanish rule, Puerto Rico was unimportant as an agricultural and mineral producer but geographically useful as a military garrison to support the expansion and maintenance of the more lucrative colonies in Mexico and the Andes. While the formal economy was tightly constrained by Spanish mercantilism, its shadow was a flourishing contraband food production sector in the 1700s and 1800s that provisioned pirates and other unregulated ships (Picó 1998, 150). Independent small-scale farming in Puerto Rico, existing on the poor, hilly farmlands of the central mountain chain and on the boundaries of mainstream society, was marginal from the beginning (Mintz 1985).

With the advent of U.S. rule in 1898, independent production became further marginalized, as the island was transformed into a platform for sugar production. Tax burdens, purposely set high, combined with the declining availability of credit and depreciated currency, forced many landowners to sell off their properties and enter the wage-labor force (Cabán 1999, 79, 173). Meanwhile, traditional European markets for Puerto Rican coffee were lost to new tariff barriers, forcing more workers into sugar. Whereas in 1897 coffee accounted for 60 percent of all Puerto Rican exports, by 1928 it made up less than 3 percent (Dietz 1986, 101). Meanwhile, sugar's proportion of export value grew from 22 percent to 53 percent between 1897 and 1930. The shift toward sugar also meant a shift toward greater trade dependence with the United States. As early as 1900, 62 percent of Puerto Rico's external trade was with the United States, up from only 19 percent two years earlier (Cabán 1999, 70).

For many Puerto Ricans, the shift toward sugar meant dependence on low seasonal wages, which condemned the majority of Puerto Rican families to desperate poverty (Dietz 1986). During the Depression, conditions went from desperate to simply untenable. Consequently, a series of strikes and *independentista* uprisings forced the U.S. government to cede more control over insular affairs to the Puerto Rican people. Puerto Rico elected its first governor, a populist hero named Luis Muñoz-Marín, in 1944. His victory took power away from a local elite that had gained wealth and prestige by collaborating with American imperialists.

Muñoz-Marín, earlier a vociferous advocate of independence, sought to strike a compromise strategy after winning the governorship in order to secure U.S. assistance for transforming Puerto Rico into a modern, independent country with a high standard of living (Picó 1998). Muñoz-Marín's administration first pursued an import-substitution strategy and built a few state-owned factories to produce goods for local construction businesses and rum factories: building supplies, glass bottles, and packaging materials. However, the anticipated customers for these state-owned firms were by and large the same elites who resented the populist victory that had displaced them from the political center. These business owners refused to purchase the products of the state-owned plants despite their high quality and low prices (Dietz 1986). The populists abandoned the project within a few years.

Next, the populist government adopted a model of "industrialization by invitation," in which U.S. firms were invited to open manufacturing plants in Puerto Rico and operate free of Puerto Rican or U.S. federal taxes for ten years. Called "Operation Bootstrap" in English and Operación Manos a la Obra (Operation Hands to Work) in Spanish, the program was designed to attract foreign firms that would create well-paid employment for Puerto Ricans while supporting the growth of Puerto Rican–owned firms by providing inputs and markets. In 1947, the first year of the program, nine companies established plants in Puerto Rico, and the number of new factories increased every year until 1953, when eighty-three new plants were built (Dietz 1986, 211). The Puerto Rican and U.S. governments have repeatedly extended the tax holiday, and it is only now being phased out.

Operation Bootstrap, however, neither ameliorated the persistent unemployment crisis nor spurred the growth of a real Puerto Rican industrial sector. The new plants did not absorb nearly as many workers as anticipated. As Dietz (1986, 2003) explains, the biggest crack in Puerto Rico's economic foundation was unemployment. The official jobless rate in Puerto Rico never fell below double digits, and the percentage of adults participating in the labor force dropped between 1940 and 1980 from 52 percent to only 44 percent. The percentage of workers employed in manufacturing grew only slightly, from 23 to 25 percent. In 1980 the nonworking population (combining the official unemployed with those not seeking jobs) accounted for 64 percent of Puerto Rican adults. Among those who worked, 30 percent worked for the Puerto Rican government.

The problems with Operation Bootstrap only worsened in the 1960s, when the government began recruiting capital-intensive industries rather than labor-intensive ones in order to avoid competition from lower-wage islands in the Caribbean (Dietz 1986; Weisskoff 1985). Meanwhile, the American companies with plants in Puerto Rico soon learned accounting methods to transfer profits earned in lower-wage off-shore locations to their Puerto Rican plants to avoid taxes, reducing the actual productive role of these plants (Weisskoff 1985). The terms of the program, then, carried little incentive to expand employment in Puerto Rico.

The Puerto Rican manufacturing sector atrophied, in large part because Puerto Rican manufacturing plants were not eligible for tax holidays under

Operation Bootstrap (Weisskoff 1985). Puerto Rican capitalists instead specialized in consumption-oriented economic sectors like importation, distribution, and retailing. Data on the three hundred largest locally owned firms in Puerto Rico (Caribbean Business 2000) show that in 1999 more than 60 percent of Puerto Rican–owned firms made their fortunes from one of four activities: (1) development (including design, construction, or the sale of construction equipment and materials); (2) the import and sale of food (primarily from the United States); (3) the import, sale, and service of cars and trucks (primarily from the United States); and (4) the import, sale, and service of other consumer goods. The "development" category accounts for the most firms out of the top three hundred (seventy firms, or 23 percent), but, tellingly, food import and retailing account for the most annual revenue (24 percent). One of the unintended consequences of Operation Bootstrap was the creation of a business sector engaged primarily in the importation of consumer goods.

Given the failure of the industrialization program to ameliorate the employment crisis, how did consumption-based businesses become so successful? Four responses to the unemployment crisis led to increased consumption power in Puerto Rico. First, unemployment was contained somewhat by the migration of Puerto Ricans to New York and other U.S. cities. With travel and logistical assistance from the Puerto Rican government, 461,000 people emigrated to the United States between 1950 and 1959, a number that exceeds one-third of the 1950 adult population (Dietz 1986). Currently, about as many Puerto Ricans live in the United States as live in Puerto Rico. Second, credit for consumers and developers became abundant thanks to a 1976 refinement of the industrialization-by-invitation policies that encouraged U.S. firms to leave their profits in Puerto Rican banks for at least five years (Dietz 2003, 143). This influx of capital fueled precisely the kinds of businesses that Puerto Rican capitalists pursued: construction and development (especially tract houses and shopping centers) and retailing of expensive items like cars and trucks and appliances. Third, as Dietz (1986) and Weisskoff (1985) surmise, the drop in labor force participation also indicates a growth in the informal economy, in both legal and illegal products and services. For example, by the early 1980s Puerto Rico had become a major player in Caribbean drug trafficking and money laundering (Montalvo-Barbot 1997).

While emigration, consumer credit, and drop in labor force participation served to ameliorate the unemployment crisis, a fourth factor contributed more directly to creating the dependent food system: the influx of government transfer payments. In 1950 transfer payments from the U.S. and Puerto

Rican governments represented 12 percent of personal income; by 1980 that proportion had increased to 30 percent. Many of those funds were from the food stamp program, in which almost 60 percent of Puerto Ricans participated (Dietz 1986, 297–99). By 1989 federal nutritional assistance as a whole accounted for more than one-fifth of Puerto Rico's gross national product (Weisskoff 1985).

What is important about these transfer payments is that they closed a circle of economic flows that incorporate Puerto Rico into the U.S. economic sphere in a mostly consumptive role. Weisskoff summarizes the structure of the modern Puerto Rican economy as follows:

> In short, the U.S. public underwrites the Puerto Rican people, while U.S. corporations shift profits through their Puerto Rican plants and back to the United States, tax free. The Puerto Rican family then buys its consumption needs, which consists for the most part of imported goods, shifting its public grant money back to the U.S. private sector. As its own economy decomposes, Puerto Rico becomes the revolving door for funds flowing from the American public back to the American corporation.... Meanwhile, these corporations employ only a token Puerto Rican workforce. (1985, 59)

In other words, the industrialization program indeed brought wealth and some measure of consumption power to Puerto Rico, but on terms that make the island a consumer but not a producer. This basic structure of the economy is reflected in the contemporary food system, one in which local agriculture has a very small role.

In the wake of sugar's ultimate decline after World War II, no other agricultural sector emerged as a new engine of the agricultural economy, and, as a result, agriculture continued its slow decline. Between 1950 and 1980, the total number of farms declined from about 54,000 to only 32,000, and the percentage of all land in farms dropped from 82 to 48. In the same period, local agriculture as a source of all food consumption fell from 51 percent to 13 percent, and the share of imports rose from 37 percent to 46 percent (Dietz 1986; Weiskoff 1985). Currently, Puerto Rico is almost wholly dependent on imported foods, and almost all of these food imports come from the United States (Benson-Arias 1997).

While food stamps played a role in weakening local agriculture, they also, paradoxically, enabled the marginalized agriculture that emerged (Carro-Figueroa and Alamo-Gonzalez 1997). Food stamps enable rural families to maintain a farming lifestyle without actually making a living from farming. More than half of the existing farms in Puerto Rico are smaller than twenty acres and more than half sell less than $5,000 worth of agricultural products annually. Recent research by Droz-Lube (2002) and Pluke and Guptill (2005) suggests that most farmers farming twenty acres or less get more income from food stamps than from agricultural sales. According to the 1997 Census of Agriculture, only 4,672 of all farmers in Puerto Rico reported earning 75 percent of their income from farming, which includes incentives and subsidies as well as agricultural sales. As of the 2000 economic census, agriculture accounted for less than 1 percent of Puerto Rico's gross domestic product (GDP) and about 2 percent of all employment. The government, in comparison, accounted for 8 percent of GDP and 21 percent of employment.

In sum, the contemporary Puerto Rican food system is defined by dependence on imports and the nutritional assistance programs that pay much of the food bill. That dependence reflects the broader structure of Puerto Rican economy and politics. Consumption, and the retail-oriented businesses that promote it, dominate the economy and wield considerable political influence. Meanwhile, Puerto Rico's fragile agrarian heritage, in which independent farming has always been a small and marginal part of the economy, was made even weaker by the deepened dependence on imports created by the industrialization program. Farming as a livelihood today has little to do with agricultural sales. Most farmers subsist on a combination of agricultural incentives, which usually require the use of particular agrochemicals, and general government transfers. Alternative agriculturalists must not only generate a whole new system of production and marketing for non-chemical-intensive agriculture, they also face a broader economy in which production itself is systematically marginalized. As the next part of this chapter explains, the structure of the contemporary Puerto Rican food system poses formidable challenges to the alternative agriculture movement.

Limits and Opportunities for Agrifood Alternatives

This section describes four significant efforts to promote ecological alternatives in agriculture. These cases reveal three ways that a food system characterized by consumption without production poses barriers to different kinds of social and economic change. First, the Puerto Rican agrifood system has created a local politics in which the powerful have a huge stake in maintaining the consumption system that fuels their enterprises. In one case, an agriculturally based community organization effort was beset by sabotage and violence from local and insular development interests. Second, while the global growth of organics offers promising investment opportunities for Puerto Rican farming, the marginal role of farmers in the broader food system gives them little incentive to invest in productive agriculture of any kind. In the second case described below, the lack of farmer investment contributed to the downfall of a promising organics exporting company. Third, because farming in Puerto Rico is more a cultural category than an economic one, activists and farmers struggle especially hard to find common strategic ground, an obstacle that impeded a recent effort to organize an umbrella association. Individually, these three cases can be explained in terms of strategic mistakes or simply bad luck. Together, however, they are intriguingly consistent with the notion that the marginal status of agriculture in Puerto Rico bespeaks a sociopolitical landscape that marks avenues for agency, not some kind of blank agrifood slate. The relative success of the fourth case described below can also be plausibly interpreted according to how Puerto Rico's particular agrifood context conditions possibilities for change.

Violent Response to Sustainable Agriculture

In the early 1980s, two experienced community organizers moved to a farm in the small community of Rabanal to begin a sustained community organizing effort. In 1984 the Association of Small Farmers of Rabanal (APARI is its acronym in Spanish) was incorporated, with Miguel A. Delgado-Ramos as executive director. APARI organized educational programs (both basic and agricultural education), infrastructure improvements for the community, and a project to grow and market poinsettias and other ornamental plants, both for entrepreneurial demonstration and to earn income for the organization.

Continuing urban expansion and relaxed land-use laws had recently accelerated the conversion of agricultural lands in the area to housing and shopping centers, in some cases in violation of laws to protect water and other natural resources. APARI led the struggle to enforce environmental laws and preserve and enable farmers' access to agricultural lands, provoking the wrath of local developers who wanted this terrain for development. Recognizing the need for affordable housing in the area, APARI also began pursuing funds for an ecological housing project, which would compete with local developers. These developers were prominent members of the conservative New

Progressive Party, which at the time controlled both the insular and local governments. Because of its visibility and initial success, APARI was met with a concerted campaign to close the organization and drive the organizers off their farm and out of the community.

Delgado-Ramos's description of the violence directed at APARI and the attempts to disrupt all the major projects of the organization is chilling:

> Things changed, because of the problems we had here with neighbors and developers that wanted to take away our earnings and get us off the farm. [They] began a project; a process of communication with other sectors, with government agencies, taking advantage of the fact that the central government and municipal government were of the same party. . . . They [began] a systemic persecution against us, against the organization, against the board of directors, against our institution. And it happened that in December of 1994, in which we were going to have a tremendous harvest of poinsettias—ten thousand poinsettias—they put herbicides in the water tank and all of our agricultural production was lost. . . . In another year . . . they cut off all of the funding from the Department of Education for which we had already been approved, $465,000, and from the Department of Housing, which had approved for us some $500,000 for our housing project. . . . [They visited] all of the members of our board of directors, threatening them that they would be harmed if they didn't leave the board. It hasn't been easy, this situation. The closed all of our sources of funding. We had a huge financial crisis. They wouldn't let us establish any agricultural production. In any moment that we left the farm . . . they would destroy all that we had with herbicides. So, it's been a difficult time. . . . We're tired. (My translation)

In addition to poisoning crops and scuttling grant-funded projects, opponents of APARI also killed horses and dogs on the farm.

Despite this campaign of violence, APARI persisted. At the time of my primary fieldwork (2000–2001), APARI was organizing a farmers' market and accompanying restaurant, beginning construction on the affordable ecological housing project, organizing a watershed protection network, continuing to sponsor general and agricultural workshops, and building agricultural production facilities (greenhouses and more) for use by landless local growers. Health issues and a political changeover at the insular level delayed these efforts up to 2005, but APARI still exists. Incidences of sabotage have eased but not entirely disappeared.

Limits of a Commercial Approach to Organic Agricultural Development

The kind of overt violence APARI experienced was relatively rare among alternative agriculture activists. More subtle and pervasive were the many barriers they faced in trying to turn a marginal and unproductive agriculture into one that is both ecologically and economically vibrant. In 1980 Jorge Gaskins, a North American who had moved to Puerto Rico about a decade earlier, initiated the establishment of the Federation of Agricultural Associations (the Fed). The Fed sought to unite regional farmers' groups to pursue grassroots agricultural development projects, linking production and processing in markets that farmers would control. In another interview (see Benedetti 1996), Gaskins explained the need for a grassroots organization focused on production in Puerto Rico: "To produce in Puerto Rico is a revolutionary act, a liberatory act, because the whole system is against it. The system is designed to make you a vegetable. If you don't produce, you don't confront our economic system and you actually support it, you become involved in maintaining it." By linking processing and marketing more closely to production, the Fed hoped to re-create a production chain driven by commerce rather than by government incentives.

The Fed's signature project was the development of marketable products (juice and pulp) from the strongly flavored passion fruit. The group secured private charitable financing to establish a processing plant that was incorporated as a worker-owned cooperative. Initially the processing operation paid good prices for passion fruit, reflecting its success in marketing the new products, but soon the influx of cheap subsidized passion fruit from Colombia and other countries undermined the market. Other companies began processing the cheap imported passion fruit and selling juice at much lower prices. The Puerto Rican government was legally unable to protect the local industry from these imports. The Fed was forced to lower its price to farmers in order to maintain the viability of the plant, and as a result farmers began to distrust the processing side of the cooperative. The project declared bankruptcy in 1985.

While the Fed did not last, the social networks and small-scale processing plant created through the Fed enabled the establishment of the most successful effort to develop organic agriculture to date. In 1991 Gaskins was approached by North American investors interested in purchasing organic and other high-quality agricultural products from Puerto Rico to sell in U.S. and European markets. According to Gaskins, the international demand for such products was obvious: "We went to three trade shows. . . . And we easily saw that there was a market for tens of millions of dollars' worth of organic tropical products. Immediately, I mean, right away. The problem was doing it. The problem really wasn't selling it. And it never was a problem of selling it." The investors founded a company called Tropical Sources in 1992 and made Gaskins a partner. The company bought and retrofitted the plant built by the Fed and hired some former Fed participants to manage the purchase and processing of organic banana, mango, soursop, tamarind, and other minor crops. Tropical sources sold fruit pulp to companies like Earth's Best, Häagen Dazs, Dannon, Welch's, Stonybrook Farms, and several European companies, especially baby food manufacturers. Within twelve months of operation, Tropical Sources had become the second-largest agricultural exporter in Puerto Rico.

To secure a supply for their organic products, the company organized and paid for the organic certification of small sections of more than fifty farms in Puerto Rico. The great majority of them were conventional farms with small, marginal areas of fruit trees that were certifiable as organic because they had been largely ignored in the farming operation and spared from agricultural chemicals. Sympathetic urban middle-class professionals with hobby farms that were similarly left out of chemical-intensive agriculture constituted another smaller group of growers. Of all the suppliers for Tropical Sources, fewer than five were active career growers invested in organic techniques.

Just as food stamps both enable and limit sub-subsistence-level farming in Puerto Rico, Tropical Sources was both enabled and limited by the conditions of the Puerto Rican food system. On one hand, Tropical Sources was based on a clever strategy: building an alternative commercial agriculture in the small farming spaces left out of the conventional chemical-intensive global food system. As Gaskins explains, "You couldn't do Tropical Sources in Mexico or Central America" because there would not be the ample chemical-free land with fruit trees or a processing plant small enough to serve a start-up organic business. The large export agroindustries in those countries consume more of the agricultural resources and demand a larger scale of processing. While Puerto Rico has unique constraints, it also, it seems, has unique opportunities.

On the other hand, while it was relatively easy to establish Tropical Sources on the margins of Puerto Rican agriculture, it was very hard to grow it from there. True to their predictions, the marketing was easy, but securing supply was the challenge. Gaskins explains: "Our problem was that Tropical Sources depended upon convincing farmers that this was going to be the best market since sliced bread. Because when they cannot do anything and pick up a better price for their product, that was wonderful, but when you wanted them to actually make a commitment to doing something proactively, that was a hard sell." Alternative agriculturalists in Puerto Rico often complain that conventional farmers just farm for the incentives, unburdened by any commitment to stewardship of natural resources, entrepreneurship, or national pride. While this critique is probably overblown, it is true that most farmers in Puerto Rico earn very little money from agricultural sales and have few choices for marketing. Under such conditions it is hardly rational to invest resources in a new kind of venture led by socially distant people and organizations.

The Tropical Sources partners expected that once the viability of organic marketing was demonstrated, more farmers would grow organically and current organic growers would use their earnings to expand production. The principals of Tropical Sources also anticipated an institutional response to their initial success. Gaskins explains, "We were taking money directly from our bottom line and investing it into organic agriculture in Puerto Rico, and we thought that that would eventually be done, that that wouldn't be a constant cost for the company. We couldn't afford it."

From the beginning, the investors in Tropical Sources intended to grow the company to a certain size and then sell it to a larger food company and realize a profit on their investment. Despite the company's initial success, they were impatient with the lack of interest in organics on the part of farmers and the Puerto Rican government. Soon the company's customers began demanding quantities of products that were impossible to supply without a significant increase in output by farmers. In trying to maximize the capacity of the plant, quality control suffered, costing the company some important accounts. When damage from Hurricane Hortense in 1995 caused the price of bananas to triple, the company closed.

Tropical Sources did not have to entice farmers to convert production to organics; rather, it had to inspire clients of the government incentive system to strike out on an unprecedented entrepreneurial path. A foreign company was not well placed to inspire that kind of shift. Some participating growers found it suspicious that the company kept the documentation of each farm's organic certification. After the company closed, some suspected Gaskins of mismanagement and financial improprieties. One distrustful grower explained his misgivings about Gaskins by saying, "He's clever, he's well connected, he speaks English," characteristics that would be considered assets under conditions of trust. Indeed, many activists admire Gaskins for these same qualities.

To understand the controversy surrounding Gaskins and the readiness of some to accuse him of such serious misdeeds, it is important to remember that few farmers (indeed, few people) are familiar with the public and private financing with which Gaskins works or the agricultural marketing process as a whole. The structure of the Puerto Rican food system largely displaces those entrepreneurial functions to remote locations. At the same time, because farmers do not need to earn much in agricultural sales to get by, they are not disposed to understand or be sympathetic about the complex and shifting difficulties that agrifood businesses face. For many, that whole aspect of the food system is absolutely opaque. As the story of Tropical Sources shows, that opacity constitutes a major barrier to collaborating for change. The next case shows that the same barrier also impedes local efforts to organize.

Divisions in Contemporary Efforts to Develop Organics

In the spring of 2001, Ismael Rios, proprietor of the largest ecological farming operation on the island, sought to organize an official association of organic farmers in Puerto Rico. His goal was to take advantage of the recent shift in agricultural development policy, after the elections of 2000 restored the Popular Democratic Party to power. The new secretary of agriculture, Fernando Toledo, vowed to increase agricultural production by 20 percent in four years through a restructured incentive and subsidy system that would encourage small-scale farmers to organize and market collectively. As a long-standing member of the Popular Democratic Party, Rios was well placed to advocate for organics before the new administration. He got Secretary Toledo to attend the first two meetings of the association to discuss coming policy changes and voice his support for organics.

The first meeting began with a blockbusting seventy people in attendance. Fifteen were leading activists from other organizations and most of the others were farmers from the area, most of whom were not at all familiar with organic farming, certification, and marketing. After some comments and discussion with the secretary, the group broke for lunch. When Rios called to order the first business meeting of the association after lunch, Secretary Toledo and all but seven of the area farmers had left. All of the fifteen activists stayed for the afternoon session, but only four in that group were career farmers.

The first task of the afternoon was to come up with a name for the organization, which prompted a brief debate between two career farmers. It began when a university professor-turned-full-time-farmer voiced her hope that the organization would include growers like her, who did not identify as organic but had similar interests in research, policy, and marketing for ecologically grown foods. While she embraced the basic principles of organic agriculture, she had doubts about the applicability of organic standards to Puerto Rico's humid subtropical conditions. For her, the notion of "sustainability," combining ecological, economic, and social concerns, was a better framework for addressing the problems of the Puerto Rican food system. Another farmer, Edgardo Alvarado, the president of the Boricua Organization for Eco-Organic Agriculture (Boricua for short), expressed concern that "sustainable" farming was "impure" because it permitted the use of agrochemicals. Advocating a "purist" position, he hoped the association would "stand firm" for chemical-free agriculture. Rios advocated inclusiveness, and the group accordingly settled on the name the Association of Organic Agriculture and Other Alternative Agricultures.

The issue seemed to be resolved, but at the second meeting of the association it became clear that the debate had only begun. On Rios's invitation, Alvarado offered the opening prayer, which included a request that God help the group "stand firm" for a "pure, clean agriculture." Rios immediately followed with a plea for cooperation: "Puerto Rico has the most important natural resources: land and people, especially farmers. Let us unite these things: the land, the farmers, and the changing philosophy in public policy. We believe in agriculture, self-sufficiency, and nature, and we can unite these things without leaving anyone behind. We've arrived at a historic moment in which these three things come together, and the government is poised to help us" (paraphrased from field notes).

Rios then described the agenda of the meeting and introduced the first item. He was interrupted by a career farmer, a neighbor and close friend of Alvarado, who began, "I want to know if this association is for organic or sustainable agriculture."

The sustainable farmer interjected, "Both have to do with ecological agriculture, and so we're on the same side."

Rios sought to forestall the debate: "We're speaking of semantics. . . . We have to be inclusive because it's an integral part of . . ."

The purist farmer interrupted again: "Sustainable agriculture permits the use of chemicals, and this does not fit into organic agriculture."

Rios tried again: "Let me continue with the report." He then spoke of a number of recent initiatives, including a legislative bill to create an organic certifying agency for Puerto Rico to comply with the U.S. Organic Food Production Act of 1990.

Soon the purist interrupted again. If we define this organization as "organic" *and* "sustainable," he protested, that "opens the door" to chemicals. He asked, "Will chemicals be a part of this organization or not?"

The sustainable farmer, sensing the implicit criticism, rose to respond. "It's clear that any organic farmer that uses chemicals loses his or her certification, and that organic agriculture has to comply with a million things. But the rules of organic agriculture are from the U.S. All of the alternative agricultures have the purpose of protecting the environment." She concluded by reiterating her argument for inclusiveness. Three people were waiting to speak, but Rios closed off the swelling debate to give the floor to Secretary Toledo.

After some comments from the secretary, the debate broke out anew. Another occasional activist, a youthful octogenarian who retired to her native Puerto Rico after a retail career in the United States, renewed the debate. "I am not a farmer," she began, "but rather an individual who wants to preserve nature, and I wish Puerto Rico were all organic. I am concerned that a person . . . stood up and defended the use of controlled herbicides."

The sustainable farmer responded, speaking directly to her critic. She explained that she'd never identified her farming as organic, described her efforts to minimize the use of chemicals, and repeated her point that organic standards already define acceptable and unacceptable practices, so her membership in the organization could not possibly dilute organic standards.

The other activist replied, "But let's be careful about connecting the word 'sustainable' to that kind of farming, because it does not work." Her comment was met with spontaneous applause from four supporters.

The meeting began to unravel. Rios asked the group to return its attention to the secretary so that he could finish his comments. Secretary Toledo fielded questions and complaints for another fifteen minutes or so until the meeting broke for lunch. During the break, the antagonists in the debate immediately grouped to continue their discussion in an animated but polite tone. Rios worried aloud that the debate was making a bad impression on the secretary and would impede efforts to organize and advocate for more public investment in sustainable alternatives in agriculture. There has not been another public meeting of the association since.

This debate is not trivial; it reveals a broader fracture in the social structure of the movement. The purists in the debate were remarkably uninterested in the secretary of agriculture (the person most able to shift resources of the state toward their stated goals), the organic marketing initiatives under way, and the nonchemical aspects of organic agriculture, like soil amendments and erosion control. The debate was not really about "organic" versus "sustainable" agriculture but rather "purity" versus "compromise." The purists reject the basic assumption behind the sustainable farmer's defense: that responsible farming involves thoughtful compromise among competing values. In other words, they spoke not as producers but as moral arbiters. Even nonfarmers, some of whom do not even garden, felt justified in insisting that the farmers' association affirm complete agrochemical abstinence. In this way the marginalization of farming has led to a situation where abstract consumer concerns rather than pragmatic producer negotiations become the basis for debates about standards. In this case, the lack of production has created conditions that impede the formation of an organization that could promote more production. The organic-sustainable debate has continued, but in the context of a newer effort: the Madre Tierra Organic Cooperative.

The Mother Earth Cooperative and Organic Farm Market: Finding Common Ground?

The Madre Tierra Organic Cooperative was founded by Adelita Rosa-García as part of a spiritual journey. She belongs to a Taino-style "Council of Elders" that has come to understand its mission as "the defense of and love for our Mother Earth." Through worship and discussion, she explains, "we began to see opportunities to organize ourselves to help in the healing of our Mother Earth and at the same time ourselves." Rosa-García's spiritual concerns are expressed in social and material terms. She seeks to use the distribution of organic foods to connect people to one another and to the land. As the cooperative recruited members and completed the steps for incorporation, Rosa-García and others began planning a farmers' market. Puerto Rico's League of Cooperatives helped them identify a small city plaza in San Juan and secure permission from the local neighborhood association to have the market there. The first market took place in April 2001.

By all accounts, the market was a success from the beginning. Vendors at the first market included five farmers (some of whom were participants in the debate described above), three stands selling prepared food, the cooperative itself, which sold dry staples like rice and oats, and two vendors selling soaps and essential oils. There was a constant stream of customers throughout the four hours of the market, and at least one farmer sold nearly all her produce. The market took place on the third Sunday of every month until 2002, when it became bimonthly. It continues to be well attended, and the number of vendors has grown to between fifteen and twenty on a typical market day. The organization and market are both called "organic," which opens the door to the debate that sidetracked Rios's organizing efforts. At early planning meetings, some participants worried aloud about the presence of explicitly nonorganic vendors, but the small number of potential participating farmers constituted a powerful argument for inclusiveness. In the spring of 2001, Madre Tierra forged a compromise: vendors were labeled with small color-coded signs to indicate whether their produce was "organic" (though uncertified), "in transition," or "sustainable." By 2004, however, the cooperative had instituted a "preinspection" program that farmers were required to complete in order to participate in the market. About twenty farmers have been inspected, and four have failed, including some who had been participating in the market from the beginning. This and other issues have spawned some conflicts within the cooperative, which has led to the resignation of some deeply involved members.

Early research on direct-marketing schemes like farmers' markets tended to see in them the seeds of a radically new food system, but recent research has been more tempered. Hinrichs (2000), for example, notes that while farmers and shoppers interact directly at the market, their relationship is usually still that of vendor and shopper. From that perspective, farmers' markets do not necessarily forge the close relationships between production and consumption that could potentially enact broadscale change in the agrifood system. In the case of Madre Tierra, however, it is precisely these modest impacts that enabled the market to begin, despite some important differences of opinion among the network of activists. The initial structure of the market, represented by the array of proximate but independent stands, provided a low-stakes but concrete basis for collaboration between producers and consumers, farmers and craftspersons, careerists and hobbyists, pragmatists and purists. With the new firmer standards, Madre Tierra will probably not be the leading umbrella project for all alternative agriculture in Puerto Rico, but the shared experience of the market may inspire more projects based on different philosophies and strategies.

Agency and Structure in the Puerto Rican Agrifood System

The case of alternative agriculture efforts in Puerto Rico suggests that the structure of the agrifood system, in this case characterized by dependence on food imports and federal transfer programs to pay for imported foods, shapes both opportunities for and constraints on agency. These findings add support to Jussaume and Kondoh's argument (see next chapter) that local context is critically important in understanding both how agency is manifest and the outcomes of efforts for change. These findings suggest that activists in Puerto Rico must contend with powerful interests in the import-dependent agrifood system that are actively hostile to projects seeking to maintain agricultural land and regenerate the working landscape. Meanwhile, the alternative agriculture movement itself is dominated by activists with a passion for agrifood alternatives but, typically, little or no experience with the complexities and contradictions of farming ecologically in the humid subtropics. Production-oriented projects like Tropical Sources are hard to begin and sustain even under favorable market conditions, and in coalition-building projects moral debates tend to supplant strategic ones. Overall, it seems that activists seeking to create a new Puerto Rican agrifood system face a particularly onerous task.

This interpretation of the Puerto Rican case thus confronts the key question addressed throughout this volume: what counts as agency? Friedland (Chapter 2) defines agency as "a manifestation of dissatisfaction with conditions of everyday life and a search for solutions" and then assesses the significance of cases of agency in terms of whether they are latent or explicit in form, individual or collective in scope, and defensive or actively resistant in approach. From this perspective, agency that does not mount a serious challenge to the status quo seems to be, by definition, a *lesser kind* of agency. Applying Friedland's framework to Puerto Rico, it is clear that the activists have expressed fundamental dissatisfaction with the status quo and have collectively pursued explicitly transformative strategies, but organic or sustainable production has not grown appreciably. Some new alternative farmers have entered into production between 2001 and 2005, but in the same period some of the most productive farmers have scaled back significantly or gone out of production altogether. These activists have clearly exercised agency, but how can we assess its significance or transformative potential when the impacts of their efforts are not yet (and may never be) tangible? What counts as success?

The results here support Jussaume and Kondoh's contention (Chapter 9) that a useful assessment of agency must locate social action not in the actors themselves but in the articulation of agency and structure. Further, these results indicate that an understanding of agency must account for socially significant efforts that nevertheless fail to produce immediate, traceable changes in the structure of the agrifood system. In Puerto Rico, the key achievement to date seems to be the social relationships among activists that are forged and renewed through market transactions, organization meetings, and other encounters. In Weinbaum's (2004) terms it is a "successful failure," failing at the direct objective but forging the social conditions that enable sustained activism and democratization. This characterization resonates with Becker's (1998) injunction to see "things" (objects, events, and outcomes) as "people acting together." Thus the new social relations forged through activism in Puerto Rico themselves are a significant achievement meriting analysis, admiration, and celebration.

REFERENCES

Becker, Howard S. 1998. *Tricks of the Trade: How to Think About Your Research While You're Doing It.* Chicago: University of Chicago Press.

Bennedetti, María. 1996. *Sembrando y sanando en Puerto Rico: Tradiciones y visiones para un futuro verde.* Orocovis, P.R.: Verde Luz.

Benson-Arias, Jaime. 1997. "Puerto Rico: The Myth of the National Economy." In *Puerto Rican Jam: Essays on Culture and Politics,* ed. Frances Negrón-Muntaner and Ramón Grosfoguel, 77–92. Minneapolis: University of Minnesota Press.

Berman-Santana, Deborah. 2000. *Kicking Off the Bootstraps: Environment, Development, and Community Power in Puerto Rico.* Tucson: University of Arizona Press.

Cabán, Pedro A. 1999. *Constructing a Colonial People: Puerto Rico and the United States, 1898–1932.* Boulder, Colo.: Westview Press.

Caribbean Business. 2000. "The Top 300 Locally Owned Companies in Puerto Rico." www.puertoricowow.com/html/top300.asp.

Carro-Figueroa, Vivian, and Carmen Alamo-Gonzalez. 1997. "Agricultural Restructuring in the Central Region of Puerto Rico and the Changing Policy Context for Mountain Farmers in the Island: Status and Prospects of Different Types of Farming Operations." Paper presented at the annual meeting of the Rural Sociological Society, Toronto, 13–17 August.

Concepción, Carmen Milagros. 1995. "The Origins of Modern Environmental Activism in Puerto Rico in the 1960s." *International Journal of Urban and Regional Research* 19 (1): 112–28.

Dietz, James L. 1986. *Economic History of Puerto Rico: Institutional Change and Capitalist Development.* Princeton: Princeton University Press.

———. 2003. *Puerto Rico: Negotiating Development and Change.* New York: Lynne Reiner.

Droz-Lube, Edna. 2002. "Pequeños agricultores frente al nuevo milenio reconsiderando las fincas familiares." Boletín de la Estación Experimental Agrícola, University of Puerto Rico.

Guptill, Amy. 2004. "The Stricken Landscape: Alternative Agriculture in Post-Development Puerto Rico." PhD diss., Cornell University.

Hinrichs, C. Clare. 2000. "The Embeddedness of Local Food Systems: Notes on Two Types of Direct Agricultural Markets." *Journal of Rural Studies* 16 (3): 295–303.

Lappe, Frances Moore, and Anna Lappe. 2002. *Hope's Edge: The Next Diet for a Small Planet.* New York: Jeremy P. Tarcher/Putnam.

Mintz, Sidney W. 1985. "From Plantations to Peasantries in the Caribbean." In *Caribbean Contours,* ed. Sidney W. Mintz and Sally Price, 127–54. Baltimore: Johns Hopkins University Press.

Montalvo-Barbot, Alfredo. 1997. "Crime in Puerto Rico: Drug Trafficking, Money Laundering, and the Poor." *Crime and Delinquency* 43 (4): 533–47.

Picó, Fernando. 1998. *Historia general de Puerto Rico.* Río Piedras, P.R.: Ediciones Huracàn.

Pluke, Richard W. H., and Amy Guptill. 2005. "The Social, Ecological, and Farming System Constraints on Organic Crop Protection in Puerto Rico." *Organic Research* (May): 16N–25N.

Raynolds, Laura. 2000. "Re-embedding Global Agriculture: The International Organic and Fair Trade Movements." *Agriculture and Human Values* 17 (3): 297–309.

Weinbaum, Eve. 2004. *To Move a Mountain: Fighting the Global Economy in Appalachia.* New York: New Press.

Weisskoff, Richard. 1985. *Factories and Food Stamps: The Puerto Rico Model of Development.* Baltimore: Johns Hopkins University Press.

Wellman, David. 2000. "Mother Nature: One Hot Mama." *Supermarket Business* 55 (3): 39, 42–48.

Yussefi, Minou, and Helga Willer. 2002. "Organic Agriculture Worldwide 2002: Statistics and Future Prospects." Report by Stifung Oekologie and Landbau (SOEL) in collaboration with the International Federation of Organic Agriculture Movements (IFOAM). www.soel.de/inhalte/publikationen/s_74_04.pdf.

INSTRUCTOR'S RESOURCES

Key Concepts and Terms:
1. Industrialization by invitation
2. Import dependency
3. Working landscape

Discussion Questions:
1. How does the marginality of conventional agriculture in the Puerto Rican food system create both opportunities and constraints for activists interested in alternatives?
2. How is the Puerto Rican case relevant to contemporary debates about the globalization of the food system?
3. What does the case of Puerto Rico reveal about the relationship between change in the food system and change in the broader society?

Agriculture, Food, and Environment Videos:
1. *The Story of Operation Bootstrap.* Pedro Rivera and Susan Zeig, 1986 (59 minutes).
2. *Field of Dreams.* People Count, N.d. (24 minutes).
3. *Life and Debt: A Film by Stephanie Black.* Tuff Gong, 2001 (86 minutes).

Agriculture, Food, and Environment on the Internet:
1. Southern Regional Sustainable Agriculture Research and Education: www.griffin
    .peachnet.edu/sare/.
2. http://puertoricowow.com/. A site offering basic history and information on the
    island.
3. Census data on Puerto Rico: www.census.gov/csd/ia/index.html.

Additional Readings:

1. Berman-Santana, Deborah. 2000. *Kicking Off the Bootstraps: Environment, Develop-ment, and Community Power in Puerto Rico.* Tucson: University of Arizona Press.
2. Cabán, Pedro A. 1999. *Constructing a Colonial People: Puerto Rico and the United States, 1898–1932.* Boulder, Colo.: Westview Press.
3. Pantojas-Garcia, E. 1990. *Development Strategies as Ideology: Puerto Rico's Export-Led Industrialization Experience.* London: Lynne Reinner Publishers.

9

POSSIBILITIES FOR REVITALIZING LOCAL AGRICULTURE: EVIDENCE FROM FOUR COUNTIES IN WASHINGTON STATE

Raymond A. Jussaume Jr. and Kazumi Kondoh

Research in the sociology of agriculture has come a long way since the late 1970s, when recurrent "farm crises" in the United States brought populist and Marxist scholars together to develop a "new political economy of agriculture" (Friedland 1991). At that time, many U.S. farm families were being squeezed out of agriculture in spite of their best efforts to sustain their farms, which contributed to economic and social decline in many rural communities. These events inspired researchers to learn more about the structural conditions that were thought to be contributing to these crises. One of their principal findings was that the economic difficulties facing agricultural producers were attributable, in large measure, to the increasing integration of agriculture into technologically sophisticated systems that were being reorganized on a global level. (For some examples of this literature, see Bonanno 1991; Kim and Curry 1993; McMichael 1994; LeHeron 1993.)

Other scholars have criticized research in the political economy of agriculture for being deterministic and for failing to pay significant attention to how people, as part of the networks they create, exert control over their own lives. For example, actor-network theory (ANT) suggests that various actors, including nonhuman actors, develop strategies in response to social conditions that promote their own interests (Busch and Juska 1997). Research that focuses on human agency, like ANT, does not necessarily reject the findings of political-economic studies but rather emphasizes the importance of conducting research in a manner that recognizes the importance of microlevel human behaviors in processes of social change (Symes 1992). This emphasis on human action, both individually and in groups, is also thought to be important in broadening our awareness of what constitutes an actor as well as significant social change. Thus, in examining processes of change in agrifood systems, some scholars now analyze the ways in which consumers engage with and influence other participants in agrifood systems (Goodman and DuPuis 2002; Lockie 2002).

From our perspective, political-economic and agency-oriented approaches to the study of agrifood systems are not antithetical. Taken together, these literatures remind us of the problems inherent in much of contemporary social science research, particularly the long-standing emphasis on causality, as well as debates over whether social change is primarily a product of "top-down" or "bottom-up" mechanisms. Clearly, human actors are important, as they do attempt to influence other actors and contribute to the creation of social structures and social change. Existing social structures, in turn, establish parameters for the thoughts and behaviors of actors. Social organizations, structural arrangements, history, ecological conditions, and human actors are equally real and equally important subjects of inquiry. For us, the question is not which of these is most significant or has greater causal influence. Rather, our interest lies in identifying the factors that shape human behavior within specific contexts of place and time, and in investigating the evolution of actors' attitudes and behaviors within these situations. In other words, we argue that research in the sociology of agriculture should move beyond a structure-versus-agency debate and devote more effort to developing analytical approaches that examine various empirical examples of how structure and agency interact.

This is the approach we use in this chapter, which seeks to contribute to a discussion of how contemporary agrifood systems are evolving by examining the contexts and processes of change in agrifood systems that vary by place within Washington. A central goal is to investigate the possibilities for the sustainable development of local food systems, which is tied to current interest in how food systems might be democratized, perhaps by reembedding them in localities (Dahlberg 2001; Lyson and Green 1999). In other words, we address the question of whether the increased use of local food-marketing strategies in different counties in Washington is a reflection of a specific intent on the part of producers and consumers to create an alternative agrifood system that challenges the mainstream industrialized agrifood system. As noted in several other chapters in this volume, particularly Amy Guptill's, our analysis demonstrates that while human agency helps shape the structure of agrifood system change at the local level, existing local structural conditions provide varied contexts within which human agency is expressed.

Four counties in Washington—Chelan, Grant, King and Skagit—constitute the empirical setting for our analysis. Each of these localities has unique ecological, historical, and social contexts that influence the possibilities for developing alternative agrifood systems. We begin our analysis by briefly describing those contexts. We then analyze survey data from consumers and producers in each of these counties. We are particularly concerned with contrasting the attitudes and behaviors of consumers and producers with respect to local agrifood systems, and thus the degree to which these actors may or may not be actively working toward the democratization of their local agrifood systems.

Theoretical Background

One of the early accomplishments of the research in the sociology of agriculture was the development of a theoretical picture of how capitalist agriculture was evolving, particularly throughout different national settings (for example, see Newby 1983). This included analyses of how national agricultural structures fit within global structures (Friedmann 1995). Early research was thus focused on analyzing how agrifood systems vary across space and time. At the same time, these analyses were prompted by a desire to uncover options that people and communities could use to challenge the political and economic realities they faced in their daily lives—in other words, to make human agency more effective.

More recent research in the sociology of agriculture has begun to move beyond a focus on the structure of *agricultural systems* to a broader investigation of *agrifood systems.* This has been due to a theoretical recognition that consumption and production systems are interconnected (Kenney et al. 1989) via an array of institutional actors (Guptill and Wilkins 2002). It also has been a product of the desire to help agricultural producers and consumers discover alternatives to the industrialized food system (Trobe and Acott 2000; Kloppenburg et al. 2000). Thus *structural* analyses of agriculture have not been simplistically deterministic or void of interest in how change can be promoted with the food system, even if such changes do not necessarily lead to a radical restructuring of our contemporary political-economic system (Goodman and DuPuis 2002).

Nonetheless, these structural analyses have not been without their limitations. One critique is that much of this work has dichotomized nature and human societies as separate entities (Goodman 1999). Scholars, including proponents of actor-network theory, have pointed out that an appreciation of the role of nature is of particular importance in the study of agriculture (de Sousa and Busch 1998; Busch and Juska 1997; Marsden 2000; Buttel 1997). Many of these same scholars also emphasize the important role that people, particularly as part of social networks, at the local level make in response to structural conditions. At its best, research that investigates the roles of various human actors in promoting change in the agrifood system also takes into account how environmental conditions, technological change, and networking between actors interact with human agency in shaping patterns of social change (de Sousa and Busch 1998).

Clearly, the agency of human beings, the "reflexive choice that is embodied in either individuals or collectivities" (Wright and Middendorf, Introduction to this volume) is important in shaping the future of all social systems, including agrifood systems. At the same time, as Wright and Middendorf point out, structures coexist with human agency. From our perspective, it is impossible to imagine either agency or structure existing in isolation. Human agency leads to the creation of structures that in turn shape human actions. As Marsden (2000) argues, it is imperative that social scientists investigate how human agency is exercised by, and emerges through, human networks, and to explain the kinds of constraints that human actors face depending on their position in social structures. In other words, analyses of agrifood system change must pay attention to the diverse ecological and historical features of specific localities and how these features interact with the everyday actions of people. By doing so, we can demonstrate how agrifood systems can be a product of the human actions that those systems have influenced.

This is the approach we use in this analysis. We are inspired in this by structuration theory, which assumes that "a social structure does not stand distinct from human action" (Molotch, Freudenburg, and Paulsen 2000, 793). According to Giddens (1984), structuration theory involves seeing agency and structure as a duality wherein all social action involves structure, and all structure involves social action. The essence of structuration theory is the dialectical interplay of agency and structure.

In an analysis of how people are trying to promote sustainable agrifood systems in their own communities, the local ecology of place can be a crucial element of the context that constrains and enables human actions. Ecological conditions such as climate, topography, and soil types are often essential to agricultural commodity production. These conditions set the field of options for farmers in terms of the agricultural commodities they can grow as well as the length and timing of the season. In turn, ecological conditions interact with demographic factors, such as population dynamics, to create the market conditions that farmers and consumers face. The degree of isolation of a farm from urban and suburban places, as well as constraints on what commodities a farmer can produce, will influence a farmer's or rancher's marketing options. For example, in Washington, where a plurality of farms are cow-calf ranches in isolated, ecologically marginal zones, finding marketing alternatives to mainstream feedlots is extremely difficult. The need to follow phytosanitary regulations, the long distances to major population centers, and the lack of viable commodity options, limit a rancher's marketing options— the reflective choices that he or she can make. Clearly, an investigation into the evolution of local agrifood systems must incorporate an analysis of how a locality's ecology, history, and social structures shape the decisions and interactions of farmers and consumers, and how the decisions of these actors contribute to the creation of altered structural conditions. In this way we seek to explore the ways in which structures and actions interact with one another in each of the four counties and to evaluate whether consumer-producer alliances are challenging the dominant agrifood system model in Washington.

Data and Methods

Our primary objective is to depict the locally specific structural contexts in which human actors make their daily life choices and form their attitudes vis-à-vis the agrifood system. To this end we combine descriptive analyses of four Washington State counties with survey data from both consumers and producers in those counties. Our sources include archival documents, which help describe the historical and ecological context, as well as individual and group interviews that were conducted during the summer and fall of 2001 with farmers, retailers, food processors, and consumers, who were identified through a snowball sampling technique. We also conducted a self-administered mail survey to farmers in 2001 and a telephone survey with consumers in 2002. This was part of a joint research project between researchers at the University of Washington and Washington State University.

For the farmer mail survey, we used a sample of three hundred agricultural landholders in each county, which represented more than 10 percent of all agricultural producers. These respondents were drawn from a list of agricultural property owners maintained by the National Agricultural Statistics Service. We used Dillman's (2000) tailored design method for mail surveys, and, after removal of ineligible surveys, the completion rate for the survey was 47.05 percent.

We selected a household sample for a telephone survey of consumers through a random digit-dialing approach. The survey was conducted in the fall of 2002 with individuals eighteen years old or older who were most involved in food buying decisions in these households. The response rate was 23 percent of all numbers that were dialed, from a low of 21 percent in King County to a high of 25 percent in Grant County. This response rate is typical for a random digit-dial approach, as it includes phone numbers for businesses and households where residents are generally absent.

The Importance of Locality

The state of Washington lies at the extreme northwestern edge of the continental United States. It is 176,600 square kilometers in size and stretches from latitude 46° to 49° north (see Map 9.1). Because of the two mountain ranges that run north-south and separate the Pacific coastal area from the bulk of the state to the east, annual precipitation varies from 4,064 millimeters annually on the coast to 152 millimeters annually in the state's central desert. This geographic and climatic diversity contributes to an incredible diversity of ecosystems, which are interwoven with particular historical

Map 9.1   Map of Washington State with County Boundaries

forces to create conditions for the evolution of diverse, locally specific structures of agriculture. These four counties were specifically selected to capture that diversity.

Skagit County lies about sixty miles north of Seattle, the state's largest city and center of the largest metropolitan district in the Pacific Northwest. Skagit County stretches from Puget Sound in the west to the Cascade Mountains in the east. White settlers first began moving into the county in the second half of the nineteenth century to engage in mining, logging, fishing, and agriculture. Because of their proximity to a large urban area, Skagit County area farms thrived. In 1925 (see Table 9.1), the size of the average Skagit County farm was twice that of King County, but in both counties the dominant commodities were milk, eggs, and various fruits and vegetables. Many of these farms were integrated operations. Thus in 1925, of the 2,858 farms in Skagit County, 2,386 had milking cows and 2,320 had chickens. Today, however, Skagit County faces urbanization pressures from the south. This is particularly true in the western part of the county, which is characterized by flat land and is bisected by a major north–south interstate highway that allows many residents to commute to work in the Seattle metropolitan area to the south. Agriculture continues to be important in the county, but the continuing disappearance of farmland has become a major local issue.

Heading east through Skagit County over the Cascade Mountains, one eventually enters the northwest corner of Chelan County. While these two counties share a border, their climatic conditions and histories are dissimilar. The eastern border of Chelan County, where much of the population is found, is the Columbia River, which flows south from Canada through the center of the state. The eastern part of the county is also in the rain shadow of the Cascades. However, as with other counties on the west bank of the Columbia, Chelan County is graced with rivers such as the Wenatchee, which flows east from the Cascades into the Columbia River system. Early settlers, many of whom were enticed to the region by the railroads, constructed small, local irrigation systems that used this river water to develop an agriculture based on fruit tree production. One intriguing element of this development is that, from its initial stages in the late nineteenth and early twentieth centuries, the expansion of fruit production acreage in this part of Washington was done with extralocal markets in mind: "The location of an orchard will also be influenced by the character of the market, whether personal or general. The orchardist has a personal market when he sells his fruit directly to the consumer, or to retail merchants in a near-by town or

Table 9.1 Historical Evolution of Agriculture in Four Counties

	KING		SKAGIT		CHELAN		GRANT	
	1925	1997	1925	1997	1925	1997	1925	1997
Number of Farms	5,125	1,091	2,858	714	2,172	1,113	903	1,699
Acres in Farm Land	140,332	41,563	137,681	93,495	158,244	123,731	63,800	1,095,099
Average Acres (per farm)	27.4	38.0	48.2	131.0	72.9	111.0	735.1	645.0
Average Value of Land and Buildings, in thousands of dollars (per farm)	7.24	378.68	7.83	609.68	14.15	414.98	13.74	1001.30
Average Value of Land and Buildings, in thousands of dollars (per acre)	0.26	8.84	0.16	4.64	0.19	3.15	0.02	1.60

Source: USDA Agricultural Census.

city; he has a general market when he sells it to a fruit shipper or sends it to commission men in the great distribution centers. Most of our Washington fruit is now necessarily placed upon the general market.... Our distant markets in the eastern states, Europe, Alaska and the Orient, are broadening every year" (Fletcher 1902, 8–9, 14).

By 1925, of the 2,172 farms in Chelan County (Table 9.1), 1,944 had apple orchards and 1,507 had pear orchards. By 1906 Washington apples were being shipped from warehouses in Wenatchee (Chelan County) and Yakima across the mountains by rail and then via the port of Seattle to Australia, Japan, New Zealand, the United Kingdom, South Africa, Egypt, and Latin America. Apples were also shipped by rail to the east, with Chicago being a major destination and port of transshipment (Sonnenfeld, Schotzko, and Jussaume 1998).

However, the development of an irrigated agricultural system that was tied into global markets was limited to counties west of the Columbia River. In the easternmost section of the state, where rainfall is greater than in the central desert, dry-land agriculture was possible. But in the center of the state, early attempts at homesteading were fruitless. Even a summer fallow system could not succeed on the "great treeless plateau" known as the Big Bend, of which Grant County is the central part, and these unsuccessful farming attempts left the region prone to wind erosion (Hunter, Severance, and Miller 1925). This began to change with the "taming" of the Columbia River. In 1933 the first of fourteen dams to be built on the Columbia River was completed at Rock Island in Wenatchee. The grandest and most famous of these was the Grand Coulee Dam. A major Depression-era project designed to provide jobs and electricity and permit the agricultural development of the Columbia Basin, the Grand Coulee Dam was completed in 1941. The water from the dam enabled the creation, in the 1950s, of the Columbia Basin Reclamation Project, which was responsible for the irrigation of nearly half a million acres of land (Dent et al. 2001).

This led to a remarkable change in the agricultural structure of Grant County. As noted in Table 9.1, between 1925 and 1997, unlike the other three counties in our study and most rural areas in the United States, Grant County experienced an *increase* in the number of farm households and in the amount of land being tilled. In addition, while cattle ranches persist in those areas of the county not suited to irrigation, Grant County became a major producer of fruits and vegetables. The bulk of the vegetable production, including potatoes, carrots, and corn, is sold to large-scale frozen vegetable processors located in the county.

The last county selected for the study was King County, home to the state's largest city, Seattle. In 1950 the population of the Seattle area was approximately half a million people, most of whom lived in the city. By 2000, the metro Seattle area had a population of 2.5 million, or more than 40 percent of the state's population. Seattle is also the center of a growing metropolitan region, which includes the city of Tacoma to the south.

A century ago, Seattle was the major port gateway city to Alaska, as well as the center of a thriving forestry industry. Agriculture did develop in this region, which has a mild and rainy climate, primarily to supply the growing urban market. In the pre–World War II era, most farms were in the "upland" region of the county, east of Seattle, and the primary commodities were milk, eggs, and fresh produce (Ruffner et al. 1942). Fresh produce, which included strawberries, lettuce, green peas, tomatoes, and carrots, was produced on truck farms, many of which were managed by Japanese immigrants. Milk and egg sales were made through producer cooperatives, and eggs were shipped to markets as far away as New York. As the population of the Seattle area grew, particularly with the growth of defense-related industries in the 1940s and 1950s, much agricultural land was developed, and agriculture in the county began to decline.

Contemporary Issues in Washington Counties

The brief review above provides a limited depiction of how these four counties, while nearby one another, have unique histories and agricultural structures. This does not mean that farmers in these localities do not share some perceptions of the world and the challenges they face. Indeed, perhaps the most common theme in many of the interviews we conducted in Chelan, Skagit, and Grant counties was the problem of low commodity prices, which were seen as a product of overproduction and global market structures. People involved in the agrifood system are very cognizant of how political and economic structures constrain the ability of farmers to engage in their craft, as many of our respondents indicated:

> One problem with the food system is that price is the bottom line rather than having the bottom line be land stewardship, an appreciation for the environmental and social value of small-scale family farms, or for organically grown produce. (Interview with farmer in Skagit County)

The wholesale market has just really priced us out of competition; we cannot compete with the corporate farms. So that's what we are learning this year. We are going through a process of possibly downsizing, possibly hanging it up. (Interview with farmer in Chelan County)

Differences across counties are expressed in how people perceive this problem of low prices and what they think are possible responses to the problem. In Skagit County the problem of low prices is compounded by urban development pressures that are driving up the price of land, thus making it even more challenging to justify farming as an economic activity. The one saving grace is that there are more perceived opportunities for direct marketing, such as farmers' markets and community-supported agriculture, because of geographical proximity to urban populations.

Farms in Chelan County, many of which were considered large when the area was developed more than a century ago, can no longer compete in a system that is dominated by packers and large-scale orchards. Indeed, many packers are now integrating backward into orchards, some of which are hundreds of acres in size and have been developed in the nearby Columbia Basin (including Grant County) over the past twenty years. This puts Chelan County farmers into a difficult predicament. As one respondent put it, "ten acres is too much fruit to sell in the direct market," but it is too small to compete in the mainstream agrifood system. Compounding the problem is that local direct-marketing opportunities are limited in size and do not feature the price potential of west side urban markets. So farmers struggling with marketing options must decide whether to drive over the mountains to urban markets in the Greater Seattle area, an option that not all feel they can use. "I think one of the problems that all of us face is that we are all just so busy that in order to find time to actually do that [go to a distant farmers' market], it is almost impossible," said one Chelan County farmer. "Somehow it would be nice to decouple ourselves from even the system that we are involved in. And I haven't figured that out yet."

For farmers in Grant County, this option appears to be even less viable. Farms there are larger, and some commodities—particularly sweet corn, sweet peas, and potatoes—are grown under contract for food processors (Washington Agricultural Statistics Service 2002). For these farms, the options are to try and stay on the treadmill or to try to find one of the emerging "niche markets" that are becoming popular. One of these is the market for organic products. However, given the scale of agricultural production and the small population base, farmers in Grant County generally do not view organics as a social movement or as a means of making their product more attractive to consumers to whom they already market directly, but as a way of increasing the value of what they produce. Many packing houses now have "organic lines," and there is also a new packing house in Grant County that is a dedicated organic packer/shipper. Organic commodities, which are comparatively easy to produce in desert conditions, are shipped from Grant County to overseas markets (Taiwan, United Kingdom), as well as to domestic markets (Chicago, Boston, Detroit, New Jersey), often via health food store chains. For many Grant County producers, marketing extralocally is inevitable. "Grant County just produces an enormous amount of food," said one, "and it's way more than the local county could ever eat anyway, so it's always going to be for export, whether that's nationally or internationally. So there is, so right away, there is the whole global market is part of Grant County."

These, then, are the contexts, both global and county-specific, within which we analyzed producer and consumer survey data from the four counties in our study. Skagit County is a periurban county with a strong agricultural history. It was once active in the production of canned, and then processed, vegetables, and is now experiencing urban development pressures. Farmers and food activists there are increasingly oriented toward "saving farms" and teaching consumers about the importance of agriculture. Chelan County is a traditional fruit-producing region that was pioneered as part of a modern, global agrifood system. It is now being displaced by global and regional forces, and opportunities for directly engagement with consumers are limited. Grant County represents the ultimate in contemporary industrial agriculture, an intense and highly efficient system in which many actors believe they can improve their economic health only by finding new ways to link to global (niche) markets. Finally, in King County, because of urbanization, much agriculture has been displaced, but the large number of potential consumers there make it an attractive market for farmers from many parts of the state who are interested in direct marketing.

Analysis of Survey Data

Direct marketing, or the use of "personal markets," is being thought of as part of the solution to the survival of small farms, as well as affording opportunities to develop producer-consumer networks that could be used to reinvigorate local agriculture and communities. Do events in localities support such optimism, and give farmers and local consumers an opportunity to engage in action that could help transform the existing agrifood system? While the strategy of direct marketing is used in all four counties, it is impossible to say at this time whether it will lead to the development of a system that would enable farms to be economically, politically, and culturally sustainable over the long term. Through an examination of survey data collected in each of the four counties described above, we can elaborate on how the agrifood systems in each of these counties are evolving. In the analysis that follows, we move beyond a test of causal mechanisms and examine participation in face-to-face markets in each county.

We begin by examining the degree to which consumers and producers engage in direct-market transactions with one another in each of the four counties (Table 9.2). We measure direct-market contact as participation, either as a consumer or as a producer, during the appropriate season, at a roadside stand, a pick-your-own, or "u-pick," farm, a farmers' market, and/or via community-supported agriculture. We differentiate between consumers who shop at one or more of these market settings at least twice a month, and between producers who do not use these channels, who use any of these channels, and who use one or more of them as their primary means of marketing their produce.

Overall, a quarter of those who participated in our consumer survey said that they shop from farmers at one or more of these venues, while 23.67 percent of the farmers reported that they sell at least some of their commodities directly to consumers. If this reflects behavior statewide, it indicates that there is a broad degree of interaction between producers and consumers. However, only 10 percent of farmers reported that direct marketing is their primary marketing strategy.

By county, a third of Skagit consumers reported buying directly from farmers, the highest percentage of the four counties. King County consumers were the least likely to use such venues; only a fifth of King County consumers shopped via these channels, compared to a quarter of Chelan and Grant County consumers. While King County consumers were the least likely to buy directly from farmers, producers in that county were the most likely to use these channels. About a third of the King and Skagit producers who responded to our survey sell via these markets, with 26 percent of King County producers relying on direct sales as their primary marketing strategy, as compared to 15 percent of Skagit producers, 4 percent of Chelan producers, and 1 percent of Grant County producers.

In other words, in the Seattle area, in the county where agriculture has declined the most, producers are the most likely to engage in direct marketing, especially if they are small farmers. Many of these farmers earn most of their income outside farming and have comparatively small household incomes. Among the King County producers who do any kind of direct marketing at all, more than half (52 percent) derive less than a quarter of their overall household income from farming. Also, a third of King County farmers have household income of less than $50,000 per year, and a third of these engage in some form of direct marketing. Clearly, direct marketing is very important for small King County farmers. It represents an important opportunity to engage in agriculture. However, a smaller percentage of King County consumers than those in the other three counties appear to be interested in buying directly from farmers. This is intriguing, as Seattle is viewed by many producers as the best direct-market location. But large numbers of consumers mask the fact that a comparatively small percentage of King County consumers are interested in buying food directly from farmers.

Table 9.2 Participation in Direct Sales Markets in Four Counties

CONSUMERS WHO . . .

	CHELAN	GRANT	KING	SKAGIT	OVERALL AVERAGE
Do not buy at direct markets (%)	32.17	31.43	35.71	23.63	30.74
Buy occasionally at direct markets★ (%)	42.17	44.08	44.12	42.19	43.16
Buy regularly at direct markets† (%)	25.65	24.49	20.17	32.81	26.11

PRODUCERS WHO . . .

	CHELAN	GRANT	KING	SKAGIT	OVERALL AVERAGE
Do not sell at direct markets (%)	81.97	85.44	68.75	66.97	76.33
Sell at direct markets (%)	18.03	14.56	31.25	33.03	23.67
Sell primarily at direct markets‡ (%)	4.42	1.09	26.15	15.38	10.28

★ Once a month or less.
† At least twice a month in season.
‡ Percentage of all farmers in county who reported direct marketing as *primary* means of marketing products.

To further develop our appreciation of the dynamics of direct marketing, we examined the relationship between household income and the predilection to participate in these direct markets. For the entire sample of consumers, the greater the household income, the more likely an individual is to shop directly from a farmer. Generally, purveyors at direct markets cannot or do not wish to compete with supermarkets over price. This understandable interest in maximizing returns is a major attraction of direct marketing for producers, but it poses a challenge for linking producers who are looking for alternatives with citizens who are involved in community food security. Low-income consumers cannot afford to pay the higher prices that farmers need to augment their incomes. While there is a relationship between income and direct-market participation for most consumers, this was not the case in all four counties.

The exceptions we discovered were in King and Grant counties, the locations of the smallest- and largest-scale agricultural production systems, respectively, in our study. Thirty percent of Grant County consumers with an income of less than $15,000 per year buy food directly from farmers, which is double the rate for households of a similar income level in the other three counties. By contrast, King County consumers with incomes greater than $100,000 per year are half as likely as households in the other three counties with the same income level to purchase directly from farmers.

Some might wonder whether participation in these markets is linked to formal education levels, with people with more education recognizing the importance of direct marketing in sustaining communities and agriculture. In the case of consumers, we could not find much of an association. Regardless of whether the respondents had a college degree, about a fifth of King County consumer respondents, and a quarter of Grant and Skagit County residents, reported making purchases directly from producers. Only in Skagit County were college degree holders (45.8 percent) more likely than consumers without college degrees (27.8 percent) to buy directly from farmers. Interestingly, Skagit County also had a very high percentage of producers with college degrees participating in direct marketing (44.2 percent) in comparison to King County (37.8 percent), Chelan County (20.3 percent), and Grant County (13.3 percent). Grant was the only one of the four counties where college graduate farmers were *less* likely to engage in direct farming. So Skagit County, which has a reputation for one of the most active social movements in the state in preserving farmland and farm households, is the one county where higher levels of formal education are associated with greater participation in direct marketing by consumers as well as producers.

This supports the hypothesis that the extent to which direct marketing promotes active consumer-producer networks interested in an alternative agrifood system may vary by locality.

One might hypothesize that consumers who live in agricultural regions are more likely to know and support farmers. We decided to examine this proposition by comparing consumers in the four counties according to some of the factors they consider when purchasing food. We contrasted the importance they attach to organic food with the importance they attach to "keeping local farmers in business." Supporting farmers economically was viewed as "very important" by 77.5 percent of consumers in Chelan County, 74.7 percent in Grant County, 70.1 percent in Skagit County, but only 58.3 percent in King County. Conversely, buying organic was "very important" to 20.8 percent of consumers in King County, 17.7 percent in Skagit, 12.2 percent in Grant, and 11.7 percent in Chelan. When we cross-tabulated these responses, we discovered that nearly 30 percent of King County consumers who thought it was very important to help sustain farmers also thought it was important to buy organic, which was double the percentage in Chelan and Grant counties. In other words, in the desert regions of the state, where agriculture is a much more significant part of the economic base, consumers are more likely to have a profarmer attitude and not to be as concerned with popular food safety issues. In the Seattle area, there is less overall support for farmers in general, but more support for farmers among those with food safety concerns.

Just as there appears to be variation across counties in which kinds of consumers are more apt to buy directly from farmers, it also appears that direct marketing is used for different reasons, which may be associated with the size and location of a farmer's operation. More than half of all farmers in Chelan County (63.6 percent) and King County (52.2 percent) who make less than a quarter of their household income from agriculture use a direct-marketing strategy. In Grant County, however, more than half (53.3 percent) of all farmers who make more than three-quarters of their income from farming use a direct-marketing strategy. Of course, large producers dominate the agrarian landscape more in Grant County than in the other counties. Thus it would seem that in Grant County, where there is a higher percentage of low-income consumers purchasing directly from farmers, direct marketing is a strategy producers often use to diversify their income source. This can take place directly or, as in the case of one of our informants, indirectly. One Grant County farmer told us that he allowed the farm laborer who lives on his farm to market some of the produce grown on the farm at a local farmers' market. The farmer does this in order to subsidize the income of the farm laborer, which in turn stabilizes the economic situation of the farm.

Finally, we were interested in examining the willingness of consumers to pay more for locally produced foods and whether this was correlated with actual purchases of foods directly from producers. In all four counties this relationship existed, although it was stronger in King and Skagit counties and weaker in Grant County. Overall, 23.3 percent of all the consumers were willing to pay a premium of 25 percent or more for local foods. This willingness was highest in Chelan County (28.0 percent), followed by Skagit County (24.5 percent) and King County (22.4 percent), and lowest in Grant County (18.3 percent). However, of all consumers who buy directly from producers, 35.4 percent said they were willing to pay the 25 percent premium for local foods. This was highest in King County (43.7 percent), followed by Skagit County (40 percent), Chelan County (39.0 percent), and Grant County (31.1 percent). In other words, consumers who lived in Grant County were more likely to recognize the importance of agriculture in their county, but were not as willing, or perhaps as able, as King County consumers to pay more for locally grown food or organic food.

Discussion

In King County, which has the largest population of the four counties but a small and shrinking agricultural base, direct marketing is an effective strategy used by many small farmers. But the percentage of consumers in King County who participate in these markets is proportionately small and driven by middle-class consumers, many of whom are motivated by an interest in purchasing organic food. In Grant County, which is the most industrialized agricultural county in Washington and the only one of the four counties where agricultural production has increased significantly in the past few decades, a substantial number of low-income consumers reported that they buy foods at direct markets. These consumers appear to be motivated more by an interest in supporting local farmers than in purchasing organic food. It is also likely that many Grant County consumers are looking for prices lower than typically found in supermarkets. This is certainly the motivation of a very active gleaning group in Grant County that combs fields after harvests for "leftover" foods. In addition, for Grant County producers, direct marketing is a way to diversify marketing strategies, not to pull out of hierarchical marketing systems. In other words, direct marketing in Grant County appears to connect a segment of the local population to the abundance of food produced in the county.

Skagit County presents yet a different context for interactions between consumers and producers. The percentage of producers and consumers who participate in direct markets and who have attended college is comparatively high. Skagit County is also on the edge of metropolitan Seattle, which means that development pressures are strong, but there are still a significant number of agricultural producers, who also have access to more direct-marketing opportunities than their counterparts in eastern Washington. There is also an active social movement, which combines producers and consumers, to protect and promote agriculture in Skagit County. There is a similar interest in Chelan County, but farmers there face different challenges. The number of local consumers is small compared to the production potential. In addition, when Chelan County was settled by nonnative farmers approximately a century ago, it was a leader in industrial agricultural development. In recent decades, however, more large-scale agricultural developments have occurred to the east, in Grant County, thus placing competitive pressure on the now comparatively smaller farms. Given this context, Chelan County has been one of the centers in Washington of farmer angst over the recent agricultural crisis in the state.

Clearly, direct markets serve different purposes and operate differently in these four counties. In part, this is because the local agricultural structure, and the degree to which the global agrifood system is connected to that structure, is unique in each of the four counties. This presents different challenges to consumers and producers in each location. Some of these can be interpreted as opportunities, others as limitations, in part depending on how they are mediated by actors at the local level. At the same time, people in all four of the locations recognize that these challenges are influenced by the same global structural realities. It is worth emphasizing that many of the people interviewed for this study recognize that the crisis facing agriculture in Washington is driven in large measure by downward pressure on commodity prices as a result of increased global trade. Of interest here is that even among the Grant County farmers who responded to our survey, fewer than a third (32 percent) agreed with the statement "free trade agreements will help my farm operation be profitable in the long term."[1]

1. The figures for the other counties were 5.7 percent (King County), 9.4 percent (Skagit County), and 35.1 percent (Chelan County).

The human actors who live in each of the places in our study can easily see that structure exists and has a direct influence on their daily lives. These same actors do their best to cope with and influence those structural forces through their actions. But it would be a mistake to interpret these actions, or at least most of them, as "resistance" to these structural forces. This is reflected in the variety of motivations and goals that consumers and producers have for participating, or not participating, in direct-marketing opportunities. We would argue that this variation is to be expected. While this diversity reflects the interests of people in each location (e.g., the interest in purchasing healthier foods or in minimizing the risks associated with managing a farm), this diversity also creates a barrier to the strengthening and expansion of alternatives, such as "personal markets," to the global agrifood system. This is expressed in the fact that many farmers' markets in eastern Washington choose not to belong to the statewide farmers' market association. The leaders of these markets in eastern Washington believe that the association cannot respond to the unique interests and problems they face in their own localities.

One of the strengths of direct-marketing initiatives, then, is that by linking local consumers with local producers these efforts can respond directly to local needs and interests. This is something that national and multinational organizational actors cannot always do effectively. At the same time, by focusing on the local, these initiatives can find it difficult to establish broader networks or create social movements that directly challenge the power of large, external actors. It is for this reason that we conclude that the possibilities for the sustainable development of local food systems are limited. Localized marketing initiatives are certainly important for the economic livelihood of the producers who participate, and they can improve the quality of life of community residents. But supermarket and fast-food chains continue to dominate local food scenes, and they are unlikely to lose their power or influence so long as alternative food systems remain disorganized on a regional or national level.

Like Amy Guptill's chapter in this volume, which found that Puerto Rico's dependence on food imports and other societal factors have held back the development of alternative agrifood systems, we found that the shape and strength of an agrifood system varies by locality in terms of history, agricultural structures, climate, and other factors. These conditions provide varied contexts within which people can act as individuals and as groups. There are people in all four counties who are dedicated participants in the "fight over food," but the ways in which agency and structure interact are not the same in all locales. Also, these interactions of agency and structure do not always directly challenge the power of the global capitalist market mechanisms that shape local contexts. How agents and structures continue to develop in their local contexts will help determine whether local agrifood systems can have a more significant influence in the sustainable development of alternative agrifood systems at the national and global levels.

REFERENCES

Bonanno, Alessandro. 1991. "The Restructuring of the Agricultural and Food System: Social and Economic Equity in the Reshaping of the Agrarian Question and the Food Question." *Agricultural and Human Values* 8 (4): 72–82.

Busch, Lawrence, and Arunas Juska. 1997. "Beyond Political Economy: Actor Networks and the Globalization of Agriculture." *Review of International Political Economy* 4 (4): 688–708.

Buttel, Frederick H. 1997. "Some Observations on Agro-food Change and the Future of Agricultural Sustainability." In *Globalising Food: Agrarian Questions and Global Restructuring*, ed. David Goodman and Michael J. Watts, 333–44. London: Routledge.

Dahlberg, Kenneth A. 2001. "Democratizing Society and Food Systems: Or How Do We Transform Modern Structures of Power?" *Agriculture and Human Values* 18 (2): 135–51.

Dent, Frederick, Lucy Jarosz, Raymond A. Jussaume Jr., Kazumi Kondoh, Joan Qazi, and Theresa Selfa. 2001. "Local Food Networks in Washington State: An Historical Analysis of Political Ecological Change." Paper presented at the joint meetings of the Association for the Study of Food and Society and the Agriculture, Food and Human Values Society, Minneapolis, 9 June.

De Sousa, Ivan, and Lawrence Busch. 1998. "Networks and Agricultural Development: The Case of Soybean Production and Consumption in Brazil." *Rural Sociology* 63 (3): 349–71.

Dillman, Don A. 2000. *Mail and Internet Surveys: The Tailored Design Method.* New York: John Wiley.

Fletcher, Stevenson W. 1902. "Locating an Orchard in Washington." Bulletin 51 of the State College of Washington, Agricultural Experiment Station, Pullman, Washington.

Friedland, William H. 1991. "Introduction: Shaping the New Political Economy of Advanced Capitalist Agriculture." In *Towards a New Political Economy of Agriculture*, ed. William H. Friedland, Lawrence Busch, Frederick H. Buttel, and Alan Rudy, 1–34. Boulder, Colo.: Westview Press.

Friedmann, Harriet. 1995. "The International Political Economy of Food: A Global Crisis." *International Journal of Health Services* 25 (3): 511–38.

Giddens, Anthony. 1984. *The Constitution of Society: Outline of the Theory of Structuration.* Berkeley and Los Angeles: University of California Press.

Goodman, David. 1999. "Agro-Food Studies in the 'Age of Ecology': Nature, Corporeality, Bio-Politics." *Sociologia Ruralis* 39 (1): 17–38.

Goodman, David, and E. Melanie DuPuis. 2002. "Knowing Food and Growing Food: Beyond the Production-Consumption Debate in the Sociology of Agriculture." *Sociologia Ruralis* 42 (1): 5–22.

Guptill, Amy, and Jennifer L. Wilkins. 2002. "Buying into the Food System: Trends in Food Retailing in the U.S. and Implications for Local Foods." *Agriculture and Human Values* 19 (1): 39–51.

Hunter, Byron, George Severance, and Richard N. Miller. 1925. "A Review of the Agriculture of the Big Bend Country." Bulletin 192 of the State College of Washington, Agricultural Experiment Station, Pullman, Washington.

Kenney, Martin, Linda Lobao, James Curry, and W. Richard Goe. 1989. "Midwestern Agriculture in U.S. Fordism: From the New Deal to Economic Restructuring." *Sociologia Ruralis* 29 (2): 131–48.

Kim, Chul-Kyoo, and James Curry. 1993. "Fordism, Flexible Specialization, and Agriindustrial Restructuring: The Case of the U.S. Broiler Industry." *Sociologia Ruralis* 32 (2–3): 61–80.

Kloppenburg, Jack R., Jr., Sharon Lezberg, Kathy DeMaster, G. W. Stevenson, and John Hendrickson. 2000. "Tasting Food, Tasting Sustainability: Defining the Attributes of an Alternative Food System with Competent, Ordinary People." *Human Organization* 59 (2): 177–86.

LeHeron, Richard. 1993. *Globalized Agriculture.* Oxford: Pergamon Press.

Lockie, Stewart. 2002. "The Invisible Mouth: Mobilizing the Consumer in Food Production-Consumption Networks." *Sociologia Ruralis* 42 (4): 278–94.

Lyson, Thomas A., and Judy Green. 1999. "The Agricultural Marketscape: A Framework for Sustaining Agriculture and Communities in the Northeast." *Journal of Sustainable Agriculture* 15 (2–3): 133–50.

Marsden, Terry. 2000. "Food Matters and the Matter of Food: Towards a New Food Governance?" *Sociologia Ruralis* 40 (1): 20–29.

McMichael, Philip. 1994. "Agro-Food System Restructuring—Unity in Diversity." In *The Global Restructuring of Agro-Food Systems,* ed. Philip McMichael, 1–18. Ithaca: Cornell University Press.

Molotch, Harvey, William Freudenburg, and Krista E. Paulsen. 2000. "History Repeats Itself, but How? City Character, Urban Tradition, and the Accomplishment of Place." *American Sociological Review* 65 (6): 791–823.

Newby, Howard. 1983. "European Social Theory and the Agrarian Question: Towards a Sociology of Agriculture." In *Technology and Social Change in Rural Areas,* ed. G. F. Summers, 109–24. Boulder, Colo.: Westview Press.

Ruffner, Woodrow, Orlo M. Maughan, Ben H. Pubols, Earl W. Carlsen, and Lawrence C. Wheeting. 1942. "Farming Systems in King and Snohomish Counties, Washington, 1939." Bulletin 424 of the State College of Washington, Agricultural Experiment Station, Pullman, Washington.

Sonnenfeld, David, Thomas Schotzko, and Raymond A. Jussaume Jr. 1998. "The Globalization of the Washington Apple Industry: Its Evolution and Impacts." *International Journal of the Sociology of Agriculture and Food* 7: 151–80.

Symes, D. 1992. "Agriculture, the State, and Rural Society in Europe: Trends and Issues." *Sociologia Ruralis* 32 (2–3): 193–208.

Trobe, H. L. L., and Tim G. Acott. 2000. "Localising the Global Food System." *International Journal of Sustainable Development and World Ecology* 7 (4): 309–20.

Washington Agricultural Statistics Service. 2002. *Washington Agricultural Statistics.* Olympia: Washington State Department of Agriculture.

INSTRUCTOR'S RESOURCES

Key Concepts and Terms:
1. Structuration
2. Agrifood Systems
3. Locality

Discussion Questions:
1. What aspects of local history and ecology in the area where you live can be said to support or restrict the invigoration of a local agrifood system?
2. To what extent have individuals or groups shaped the history and environment of the locality where you live?
3. What extra-local government and business actions have had an impact on the viability of a local agrifood system where you live?

Agriculture, Food and Environment Videos:
1. "Beyond Organic." Bullfrog Films, 2000 (33 minutes).
2. "Broken Limbs." Bullfrog Films, 2004 (57 minutes).

Agriculture, Food and Environment on the Internet:
1. Washington State University Small Farms Team—http://smallfarms.wsu.edu/about/
2. Food Consumption in Washington State—http://crs.wsu.edu/outreach/rj/ag-consumer/index.html
3. Agriculture in Washington State—http://www.crs.wsu.edu/outreach/rj/agsurvey/index/html

Additional Readings:
1. Joan Qazi and Theresa Selfa. 2005. "The Politics of Building Alternative Agro-Food Networks in the Belly of Agro-industry." *Food, Culture, and Society*, Vol. 8 (1): 45–72.
2. Lucy Jarosz. 2000. "Understanding Agri-Food Networks as Social Relations." *Agriculture and Human Values*. 17 (3): 279–83
3. David Sonnenfeld, Thomas Schotzko, and Raymond A. Jussaume Jr. 1998. "The Globalization of the Washington Apple Industry." *International Journal of the Sociology of Agriculture and Food*. 7: 151–80

10

CONSUMERS AND CITIZENS IN THE GLOBAL AGRIFOOD SYSTEM: THE CASES OF NEW ZEALAND AND SOUTH AFRICA IN THE GLOBAL RED MEAT CHAIN

Keiko Tanaka and Elizabeth Ransom

With the rapid growth in studies of globalization and mass culture (e.g., Ritzer 1992), social studies of agriculture and food, or agrifood studies, have begun to examine social relations and practices surrounding agricultural production and food consumption in the context of rapid capital concentration in the global agrifood industry (e.g., Bonanno et al. 1994; Goodman and Watts 1997; McMichael 1998). Within the global agrifood system, however, there are tremendous variations in the ways in which actors, from farm to dinner table, are linked together. This is because biochemical and ecological characteristics of food items often, though not always, define where, when, and how these items can be grown, processed, and distributed (for example, wheat versus tomatoes). Therefore, many agrifood scholars (e.g., Bonanno and Constance 1996; Busch et al. 1991; Dixon 1999; Friedland 1984, 2001; Heffernan 1984) use commodity chain analysis as a means of conducting research on different agrifood products. By following a given food item from farm to dinner table, these scholars examine how the organizations that produce, circulate, and regulate a given commodity are arranged in society.

With few exceptions (e.g., Dixon 1999; Long and Villarreal 1998), however, many commodity chain studies tend to privilege economic relationships more than any other type of social relationship, and therefore producers and consumers are conceptualized as merely economic categories (Krippner 2001). Very little effort has been made in the literature to question the

The case study of New Zealand was funded by a Faculty Research Grant from the University of Canterbury; the case study of South Africa was funded by a Doctoral Dissertation Improvement Grant from the National Science Foundation. The authors thank Wynne Wright, Gerad Middendorf, Carmen Bain, and anonymous reviewers for valuable comments on previous drafts.

conceptualization of the categories *producers* and *consumers,* what classifies them into these categories, and who constructs these categories.

We argue that in the highly globalized agrifood system, the categories of producers and consumers are elusive because very few countries in the world are fully self-sufficient in food production. Even if self-sufficient, most countries, including the United States, rely on imports to ensure the availability of a wide range of food products year round (e.g., coffee and bananas for U.S. consumers). Moreover, farmers and food companies in some countries may rely on consumers overseas more than on consumers in their own countries for their livelihood. To understand how the global agrifood system operates, it is therefore important for us to ask, who produces our food? Who are the consumers of the foods that "our" farmers and food companies produce?

This chapter aims to show that the process of changing rules within the capitalist market system, specifically meat safety governance reform in New Zealand and South Africa, raises profound obstacles for human agency, yet opens new spaces for conceptualizing who participates in promoting change. Agency and structure are complex concepts with dueling tensions that alter the form and substance (as Wright and Middendorf argue in their Introduction to this volume) of individual and collective action in the red meat commodity chains of these two countries. We show that, far from being monolithic, the ways in which capitalism and a changing agrifood structure affect actors in a commodity chain, and the ways in which these actors respond, vary across time and space. We hope to make clear the ways in which structures affect agency, but we also aim to show how structural changes open new opportunities for agency.

The definition of agency presented in other chapters in this volume—the active reflexive choice of individuals or collectivities—requires researchers to differentiate between intended and unintended, conscious and unconscious actions. We conceptualize *agency* slightly differently, so as to allow a focus on actions and the consequences of acts. We argue that the emphasis of empirical investigation needs to be shifted from intentions and motivations to tools and mechanisms that facilitate actors' ability to act. In other words, such individual capacities need to be situated in a web of relationships that constrain or enable action. This kind of analysis exposes how various types of food networks (e.g., export-dependent versus domestic commercial networks) in a given commodity chain collide and converge—creating new allies and conflicts and simultaneously redefining the role of each actor in both local and global agrifood markets.

In this chapter, there are three interrelated but distinct types of food networks surrounding red meat: export-oriented, domestic commercial, and informal. By focusing on changes in the meat safety governance system in New Zealand and South Africa, this chapter will show the fluid, unstable, and ambiguous nature of producers and consumers as categories in these three networks. Producers construct their consumers and act to respond to the needs of these constructed consumers, not necessarily to the needs of those who actually consume their products. Thus, we will deconstruct the category of consumers in order to discuss how agency is constructed and implicated through negotiations to reform the meat safety regime. Our case studies show how each actor in the red meat chain differentiates *consumers* and *citizens* and uses these categories to justify their actions in meat safety governance reform in their nation.

Consumers Versus Citizens in Global Reform of Meat Safety

Today, a slice of beefsteak or lamb chop that most of us eat for dinner in the United States is unlikely to come from cattle or sheep raised on a family farm in our area. It may not even come from our country. This is because a relatively small number of large multinational firms in the United States, Europe, and Japan tend to dominate the meat trade in industrialized countries, controlling the distribution of meat from feedlots to grocers (Dyck and Nelson 2003). At the same time, we have begun to hear about a series of meat safety scares caused by food-borne pathogens, such as *E. coli O157:H7*, bovine spongiform encephalopathy (BSE, or mad cow disease), and avian influenza virus H5N1 (bird flu), that threaten our health and the viability of our agrifood economy. Some (see, for example, Walters 2003; Schlosser 2002) attribute an increase in these diseases to certain practices (e.g., BSE caused by feeding cows with ruminant remains) and mechanisms associated with intensive production of agriculture and concentration of food distribution. They point out that when food contamination does occur, it reaches more people in a shorter period of time (e.g., meat contaminated with *E. coli O157:H7*). What can we do to protect "our" meat from such deadly pathogens? Should our goal be public health or economic stability—or both?

CONSUMERS VS. CITIZENS

When we discuss the food system, especially creating change in the food system, we are confronted with the fact that everyone who participates in the food system is simultaneously a citizen and a consumer. Gabriel and Lang (1995, 175) recognize that the citizen is generally a political concept, "defining individuals standing within a state and a community, according them rights and responsibilities." The consumer, by contrast, is an economic concept; individuals "need not be members of a community, nor do they have to act on its behalf . . . consumers operate in impersonal markets, where they can make choices unburdened by guilt or social obligations" (174). Marsden, Flynn, and Harrison (2000, 47) show the usefulness of differentiating citizens from consumers in understanding changes in the role of the nation-state in food governance, and consequently in citizenship rights to food, as "the legitimation of social and political claims made on the part of the people." They point out that in the past two decades the traditional role of the state as regulator of both production and consumption practices has declined, as more privatized and differentiated forms of rights to food provision have increased. This change in food regulation facilitates a shift from citizens to consumers in the agrifood system.

Today, consumers are increasingly differentiated by how much they are willing to spend on food (e.g., store-brand tomato sauce sold at a national chain store versus gourmet-brand organic tomato sauce sold at a specialty food store). Often, there is little *totality* (or coherence) in the behaviors of individual and collective consumers. Many organic and local food consumers also continue to buy conventional food. Thus agrifood commodity chains have experienced a shift from producing for the assumed *citizen* to producing for multiple *consumers*.

In consumer-oriented production, certain food products are marketed using labels that identify special categories, including animal welfare, environmental sustainability, and fair trade, to name a few. Yet these issues are relevant to citizens and not merely to consumers, because they raise the question of how to create a better society (Brom 2000). On the one hand, in a global consumer culture, as Gabriel and Lang caution, *consumers* are encouraged to be atomized individuals with little interest in promoting a common good for a larger community. On the other hand, *citizens* are needed to participate in the global public debates over what constitutes a good agrifood system and to act collectively to reform institutional mechanisms for governing the existing system.

For example, after the massive outbreaks of *E. coli O157:H7* in the United States in 1993 (see Juska et al. 2000; PBS 2002) and of BSE in the United Kingdom in the early 1990s (see Draper and Green 2002; Millstone and van Zwanenberg 2002), some consumers chose to express their concerns publicly and collectively demanded changes in the nation's meat safety regulations. As Draper and Green pointed out (2002, 611) in the case of the UK, such active public engagement reconstructed "*consumers,* who could avoid risks through making informed choices," as "*citizens,* not only reacting to information about risk but also having an obligation to contribute to policy formation." The transformation of consumers into citizens affected not only consumers, who could afford to demand change through their purchasing decisions, but all citizens in the UK, regardless of differences in their purchasing power. In both the United States and the UK, after months of negotiations between these concerned citizens and key actors representing the interests of the agrifood industry and the national governments, a new framework of food safety governance emerged that relies on tools such as the hazard analysis and critical control point (HACCP) system to manage potential food safety risk. In short, this new food safety framework was the outcome of actions that express the agency of diverse actors in the agrifood system, each with different sets of economic interests, political motivations, and ideological perspectives.

HACCP AND THE REORDERING OF THE AGRIFOOD SECTOR

By the late 1990s, both government and industry actors in many industrialized, and some developing, countries had begun adopting the new framework of food safety governance, partly to respond to public concerns with food safety issues within their countries and partly to maintain their access to global agrifood trade (Roberts and Unnevehr 2003). As the HACCP-based food safety governance became incorporated as a new structural feature of the global agrifood market, new opportunities became available for some actors, while others confronted new constraints.

This is because food safety governance reform reorders the social relations and the distribution of power among actors in the agrifood system, both in a given country and between countries. A given food safety regulation classifies people (e.g., "farmers," "retailers"), organizations (e.g., "Ministry of Food," "Food Safety Inspection Service"), equipment (e.g., "meat recovery equipment"), tools (e.g., "butcher knife"), plants (e.g., "grain plants"), and animals (e.g., "meat animals") into distinctive categories, each with (re)assigned roles and responsibilities. Then, individual actors in a given classification (e.g., farmers, processors) are furthered differentiated into hierarchical groups based on their financial, technical, and moral capacity to adhere to these regulations. Food safety governance with a set of regulatory measures and procedures, such as the HACCP system, therefore becomes an important institution of discipline and surveillance (Foucault 1979) that determines who can be included and excluded from a particular market, whether local or global.

The U.S. Department of Agriculture's Pathogen Reduction/HACCP Act of 1996 exemplifies this kind of classification and differentiation, especially in terms of distributing power among both domestic and international actors. This act requires that slaughter/packinghouse premises wishing to sell products in the United States operate under their own risk-management plan (called HACCP), which identifies, monitors, and controls known biological, physical, and chemical hazards (Juska et al. 2000). While the United States has no jurisdiction over slaughter/packinghouses in other countries, most major meat-producing and trading companies outside the United States adopted the HACCP system, largely in order to retain access to the U.S. market—the world's single largest export and import market for meat. However, the HACCP requirement involves additional costs (e.g., upgrading equipment) and tasks (e.g., frequently checking water temperature) in production, and therefore raises the price of meat products and puts plants with limited resources at a disadvantage (Ollinger, Moore, and Chandran 2004).

Both New Zealand and South Africa are meat-producing countries that have been greatly affected by the enactment of the 1996 Pathogen Reduction/HACCP Act. As Table 10.1 shows, New Zealand and South Africa play very different roles in the global red meat chain. Although New Zealand ranks second in the world for mutton (mature sheep) and lamb (sheep under twelve months of age) production, compared with their competitors, the levels of beef and veal (meat from cows under three months of age) production in New Zealand and South Africa are low (FAO 2003). In 2003, for example, the world's two largest beef and veal producers, the United States and Brazil, produced 12 million and 7.4 million metric tons, respectively. In contrast, New Zealand and South Africa each produced less than 0.6 million metric tons, making them thirteenth and twelfth in the world's beef and veal production (USDA-FAS 2003).

What makes the case of New Zealand unique, however, is that approximately 80 percent of its beef is consumed in more than eighty different countries (New Zealand Meat 2003a), making it the world's fifth-largest exporter (USDA-FAS 2003); and 90 percent of the nation's lamb is consumed in more than one hundred countries (New Zealand Meat 2003b). In 2000, New Zealand exports of cattle and sheep meat, edible offal, and other animal by-products were worth NZ$4 billion, or one-seventh of the nation's total exports (NZMAF 2003). In other words, New Zealanders are not necessarily the most important consumers for the red meat industry in New Zealand.

In contrast, South Africa imported 37,000 metric tons of beef and veal (USDA-FAS 2003) and 34,192 metric tons of mutton and lamb (FAO 2003). Yet the case of South Africa illustrates opportunities and constraints in transforming the agrifood sector in a developing country. The latest projections estimate that by 2020, with rapid population growth and urbanization, 63 percent of meat consumed worldwide will be produced in developing countries (Haan et al. 2001). Higher-income developing countries with a fairly developed agricultural sector like South Africa are expected to lead livestock production in the developing world.

Although New Zealand and South Africa play very different roles in the global red meat chain, both countries have recently begun to transform the institutional framework for regulating meat safety, largely as part of the effort to harmonize with food safety regulations accepted by the United States, Canada, and European countries. More important, as Table 10.2 shows, the

Table 10.1   Beef and Veal Production, Consumption, and Trade in Selected Countries in 2003

	Production		Import		Export		Consumption	
	1,000 metric tons[1]	Rank	1,000 metric tons[1]	Rank	1,000 metric tons[1]	Rank	Kg/ person/ year[2]	Rank
Australia	2,050	6	4	22	1425	1	32.51	6
Brazil	7,430	2	80	11	970	3	37.03	4
Canada	1,255	10	340	7	615	4	31.11	7
European Union[3]	7,260	3	520	4	530	6	19.48	10
Japan	525	15	850	2	0		10.58	15
Mexico	1,950	7	500	5	8	15	23.87	9
New Zealand	665	13	10	19	535	5	35.00	5
Russian Federation	1,700	9	800	3	5	16	17.40	11
South Africa	670	12	15	15	25	13	14.41	12
United States	11,993	1	1,481	1	1163	2	42.87	3

Sources: USDA-FAS (2003); World Bank (2005).

  1. Carcass weight.
  2. Total domestic consumption was divided by total population. The 2003 population data of World Bank were used.
  3. Includes fifteen member states in 2003.

red meat chain in each country consists of multiple food networks, namely: (a) an export-dependent network, (b) a domestic commercial network, and (c) a domestic informal network. New Zealand, with a population of 3.3 million, has for decades pursued the path of export-oriented agricultural development, even though the meat industry consists of numerous relatively small-scale companies owned by private New Zealand individuals or sheep-cattle farmer cooperatives (New Zealand Meat 2003c). In South Africa, the institutionalization of apartheid effectively created and maintained a two-tier agricultural system: the commercial industry, represented by a small number of whites, and subsistence farming by blacks and "coloreds," who make up the vast majority of the country's population. Today, South Africa's government is attempting to put an end to the two-tier agricultural structure.

As discussed below, who constitutes the consumers and the producers differs significantly among these networks. In short, asymmetries between these networks make categories such as producers and consumers something that must be constantly negotiated and (re)constructed among actors. As shown in the next two sections, by negotiating the redistribution of the power to manage social order in the agrifood system, food safety governance reform becomes a site where the categories of producers, consumers, and citizens are articulated not merely as economic but also as political and social actors.

The Case of New Zealand: Collision of Global Citizens and Marginalized Consumers

Lamb, beef, and mutton, along with wool, are the major export commodities of New Zealand.[1] Consequently, the nation's red meat chain has long been dominated by actors engaged in the production and distribution of export meat products. Yet the neoliberal reform of 1984 (Le Heron and Roche 1999), which eliminated virtually all government support programs for the agricultural sector, exacerbated the uneven distribution of economic resources and political power among the actors in three distinctive types of networks—export-oriented, domestic commercial, and domestic informal—within the red meat chain. This structural feature affected, and was affected

---

1. The case study of New Zealand was conducted between January 1999 and April 2001, starting when the Animal Product Act (APA) was in the final stage of negotiations, and completed in the middle of the three-year transition period from the Meat Act of 1981 and the Apiaries Act of 1969 to the APA. Our study is based on the analysis of various documents concerning the act and on interviews with thirty-five individuals representing twenty-five organizations or groups.

Table 10.2 Three Networks of a Red Meat Chain, New Zealand vs. South Africa

	New Zealand		South Africa	
	Producers	*Consumers*	*Producers*	*Consumers*
Export-Oriented	Farmers, large-scale meat processors, and medium-size abattoirs	Overseas consumers	Large-scale white farmers, large-scale meat processors	Overseas consumers
Domestic Commercial	Farmers, small-scale abattoirs, supermarkets, butcher shops	Urbanites, foreign tourists	White farmers, meat processors, abattoirs, supermarkets, butcher shops	Upper-middle-class and middle-class urbanites
Domestic Informal	Farmers, rural dual operators	Rural residents	Black farmers, informal slaughterers	Poor

by, how the agency of New Zealand consumers was articulated in negotiations for meat safety governance reform.

In the export-dependant network, most products produced in New Zealand are shipped overseas. The label *export quality* on these products distinguishes products processed by export meat companies from small-scale, domestic-only abattoirs, whose products are often sold at traditional neighborhood butcher shops. This distinction between "export" and "domestic" meat products on the domestic market creates an apprehension among New Zealand consumers that they are left with "a spill-over from export," or only those products with "inferior quality that failed to meet export standards," as many New Zealanders have testified. The national government has always regulated the export network to ensure that the export requirements of diverse countries are met and that domestic slaughter does not infiltrate meat destined for export. Until recently, however, little effort was made to monitor the two domestic networks, either the commercial or the informal.

New Zealand established the Food Safety Authority (NZFSA) in 2002 to combine the food-related functions of the Ministries of Agriculture and Forestry and of Health and unify the authority for regulatory oversight under a single agency. Following the trend of many industrialized, and some developing, countries (Roberts and Unnevehr 2003), the NZFSA adopted the emphasis on a "farm-to-table" approach in dealing with potential hazards and the HACCP system as a basis for risk management. Thus the agency immediately took over the administration of (a) the Animal Product Act of 1999, which *requires* producers of meat products to develop and implement their own risk-management plan based on HACCP principles, and (b) the Food Act, which *encourages* retailers of meat products (e.g., supermarkets, restaurants, hotels, butcher shops) to develop and implement a food safety plan, also based on HACCP principles. These HACCP-based regulatory measures were to help the NZFSA simultaneously achieve two goals, the protection of public health and safety and the promotion of stable trade and commerce (NZFSA 2002).

How do these two new regulatory measures under the new agency, or the new meat safety regime, in New Zealand affect the three networks of the red meat chain, particularly the different types of producers and consumers implicated in the chain?

In all three of the networks, the producers refer to sheep or cattle farmers and meat-processing companies *in* New Zealand. Indeed, most farmers, meat-processing companies, and abattoirs belong to *both* the export-oriented and domestic commercial networks. The new regime required little change in the actions of producers involved in the export network, as export-oriented meat companies had been voluntarily implementing their own HACCP plans for more than a decade (Cutt 1998). According to a leader of a farmer organization, most farmers are accustomed to preparing statutory declaration forms that record how each batch of animals was treated on their farm, so as to demonstrate that their animals are "fit for purpose" (i.e., that the goods are fit for the purpose the customer wants) to be exported.

There are producers, however, though small in number, who are involved *only* in the domestic commercial or informal networks. Some farmers choose not to or are unable to participate in a quality-certification program of an export-oriented company. Similarly, there are small-scale abattoirs that specialize in slaughtering for the domestic market. In rural New Zealand (14 percent of the population live outside concentrated settlements of a thousand or more people), the informal network also continues to operate, as some rural farmers choose to sell their products through direct marketing at their roadside markets or farm-stay operations (i.e., agrotourism). Their actions have been greatly affected by the new meat safety regime. Many small abattoirs and independent retailers now have to develop their own HACCP plan and upgrade their skills and facilities to implement them (Bristow 1999). Moreover, sheep and cattle farmers are prohibited from supplying meat for recreational catch (or so-called unregulated meat) to workers or their fee-paying guests without an approved HACCP plan (*Southland Times* 2000).

These changes imposed on the producers suggest that the new meat safety regime functions as a "disciplinary institution" in which actors who fail to produce meat "good" enough for export markets are punished by losing access to the domestic market, while those who succeed are rewarded by gaining access to the export network. In short, only actors who are able to participate in the export network can remain producers. In our interviews, NZMAF officials said that further consolidation in the red meat chain over the next two decades would be "inevitable" and "necessary" to build consumers' trust in the safety of New Zealand red meat products.

If the social order created by the new meat safety regime aims to enable every producer to be part of the export-oriented network, then whose trust in meat safety does the new regime try to build?

In the export-oriented network, consumers mean customers or clients from eighty-plus countries who purchase products that contain New Zealand red meat. For the producers in this network, New Zealanders are hardly considered among the consumers, though a large portion of them consume meat products produced in this network rather than those exclusively produced in the domestic commercial or domestic informal network. Some representatives of the major meat companies in the export-oriented network justified their active support for meat safety governance reform, particularly the enactment of the Animal Products Act, by stating that this new regime would enable them to more effectively respond to food safety concerns raised by consumer organizations overseas. During interviews with meat company representatives, they downplayed the impact of lobbying activities by New Zealand consumer organizations on their businesses or the negotiations of meat safety governance reform. Yet both government and industry representatives in the export-oriented network described New Zealand consumers as the ultimate beneficiaries of this regulatory reform.

Meanwhile, producers involved only in the domestic commercial or domestic informal networks expressed less enthusiasm toward the two new regulatory measures. For them, the consumers mean New Zealanders, and the consumers are stratified according to their residence (in the country or the city) and socioeconomic capacities (whether or not they can afford premium meat products). One interviewee noted that many "dual-operator" butchers, or retailers who sell both regulated and unregulated meat, in the domestic informal network were opposed to the reform because of financial and technical constraints imposed by the HACCP requirement. According to her, they were afraid of losing their business and forcing their consumers, who reside in or visit rural New Zealand, to either travel great distances to purchase approved meat products in urban areas or lose opportunities to enjoy rural New Zealand. For some rural residents, the safety of meat products is compromised when they must drive long distances from a city with unrefrigerated meat in the car. Yet some producers in these two domestic networks suggested that urban New Zealand consumers might benefit from the new meat safety regime because their meat would be regulated the way export meat is.

As in the case of South Africa, with an increase in the market share of supermarkets in the urban areas, New Zealanders have become more conscious about food safety issues. Unlike many red meat–producing countries in the world, however, New Zealand has never experienced an outbreak of *E. coli O157:H7*. Given tight border controls on the importation of biological organisms, there have been no cases of BSE or foot-and-mouth disease (FMD), which pose public health risks and damage the industry. According to food safety researchers, New Zealand has one of the most effective surveillance systems in the world for monitoring public health–related diseases. As a leader of one consumer organization pointed out, "HACCP plans and these kinds of food safety plans often do not deal with the kind of food safety issues that [New Zealand] consumers are most concerned about" (e.g., genetic modifications, antibiotic use, pesticide residues). In other words, the constructed consumers with whom the new meat safety regime tries to build trust are not New Zealand consumers but those who reside overseas. Thus meat safety governance reform is driven not by the desire or need to protect the domestic consumers, but rather by the effort to formalize the HACCP system that has been adopted by the producers in the export-oriented network and improve their capacity to access overseas markets.

Is there, then, no agency for New Zealanders in this reform of food safety governance? Are New Zealanders merely marginalized consumers in the global red meat chain?

If we shift our attention and ask how New Zealanders might have articulated their agency as citizens, we can interpret this story from a different angle. Then meat safety governance reform in New Zealand can be examined as a *political* process involving the "public policy choices" of citizens, as well as an *economic* process involving the "purchase choices" of consumers. As pointed out earlier, however, consumers are increasingly differentiated based on their geophysical, socioeconomic, cultural, and ideological capacities to purchase food. How would citizens effectively present "the consumer's voice," given such diverse and contradictory opinions, views, and practices concerning food consumption? How would atomized consumers with little sense of social obligation transform themselves into citizens who accept responsibilities for shaping public policy on behalf of their country?

To answer these questions, let us look at a series of institutional changes that took place in the 1990s in the food safety policymaking process as the result of two sets of actions by New Zealand citizens. These actions may be interpreted as a signal, if not a direct action, of how New Zealand *citizens* responded to transformations in *their* agrifood system that in the past two decades have increasingly alienated New Zealanders by making them second-rate consumers and restricting their purchase choices of food.

The first set of actions concerns voting for national and local political candidates who consider food safety issues a priority. New Zealand has a parliamentary system similar to the British. In the 1990s, for example, the Green Party, running on a platform of environmentally sustainable development, began to see its members getting elected at the local level (Green Party of Aotearoa, New Zealand 2003). In the general election of 1999, the balance of power in Parliament shifted for the first time in ten years from conservative to liberal with the development of a coalition of Labor, Alliance, and Green parties. By winning parliamentary seats for the first time as a stand-alone party, the Green Party pushed food safety issues to the top of the nation's political agenda and led the effort to organize the Royal Commission on Genetic Modifications in 2000, a nationwide inquiry into the benefits and risks of genetic modifications of plants and animals. In a nation with a population of fewer than 4 million and in the space of twelve months, the Royal Commission conducted a public opinion survey, held fifteen public meetings, sponsored a Maori consultation process involving twenty-eight workshops and twelve *hui* (gatherings), put on a youth forum, ran a public submission process resulting in more than ten thousand written entries, and held thirteen weeks' worth of formal hearings that involved more than a hundred interested persons and approximately three hundred witnesses within and outside New Zealand (Royal Commission on Genetic Modification 2001, 6).

The second set of actions by citizens involves the expansion of consumer representation in public policymaking. In meat safety governance reform, consumer representatives were requested by government and industry representatives to participate in the decision-making process through various channels. For example, David Russell of the Consumers' Institute, which publishes the monthly *Consumers Magazine,* invited the Meat Industry Standards Council to negotiate the particulars of the Animal Product Act with government and industry representatives. All the discussion documents and policy drafts concerning the reform were made available through the Web sites of government agencies; consumers thus engaged in policy decisions and became active citizens. As an official at NZMAF pointed out in an interview in 1999, "In the last two years, I think the industry has realized that they can't hide—[New Zealand] consumers want accountability. . . . We [the NZMAF] have to be careful, not to forget *the other side*—the consumer. So we can't look like we have been siding up with industry. It would have to be a triangle of people" (emphasis added).

Furthermore, Phillida Bunkle, a member of Parliament from the Alliance Party and the minister of consumer affairs, justified the establishment of NZFSA as follows: "It's incredibly important to get this right *for New Zealand* [citizens], especially in terms of getting our quality assurance right *for our markets.* I think that whatever structure [for a new food safety agency] gets developed, then [New Zealand] consumer interests have to be more strongly represented than they are now" (emphasis added). Russell himself and other consumer activists see the need for wider representation of consumers in any legal and regulatory reforms in the agrifood sector that would inevitably affect the well-being of New Zealand consumers. As Russell pointed out, no organization "can represent consumers. So [Consumers' Institute does] not represent consumers' views, but represents consumers' interests." To give voice to the interests of New Zealand consumers in food safety policymaking, nine consumer and environmental organizations, each representing a different segment of New Zealand consumers, formed the Safe Food Coalition.

These concrete institutional changes emerged from actions by New Zealanders and effectively created a space in which individual consumer voices in the impersonal market are transformed into a coherent set of collective actions in the political sphere. Certainly, none of these actions may be interpreted as an expression of *resistance* to meat safety governance reform. The establishment of the new meat safety regime proceeded with little opposition from consumer or environmental organizations. This was because the demand of New Zealand consumers for stronger food safety governance happened to correspond with the desire of the producers in the export-oriented network to legitimize the HACCP-based system, though HACCP may not have been a priority area for consumers.

Active, collective citizenship in New Zealand faces challenges, not necessarily "from the assertion of individual consumption rights" (Marsden, Flynn, and Harrison 2000), but from the expectation that New Zealand citizens *should* support policies that would help to grow the nation's economy *and* political power in the global system. As Bunkle put it, "The only way to convince [New Zealand] consumers is to make sure that there is integration between consumer and producer interests. The current system was heavily weighted towards producer interests, but I think the two interests—producer and consumer—are one and the same. . . . We are food producers and we are all consumers. So, we had better get it together if we are going to protect New Zealand's future."

Such a view, widely shared by government and industry representatives, suggests that New Zealanders are important allies for actors in the export-oriented network, even though they may not be as important as international consumers. Their enrollment as citizens in this network is necessary to legitimize the actions of these actors to pursue their economic and political agendas in the global agrifood system. To protect the future of their nation, New Zealanders therefore are expected to exercise their agency as *citizens*, rather than as *consumers*, who select their political representatives and participate in the processes of policymaking. By doing so, they are, though indirectly, shaping how various actors in the New Zealand red meat chain participate in the global market and what kind of images about the nation these actors present to the world.

The Case of South Africa: Building the Modern Food System

If the New Zealand red meat chain illustrates how globalization forces are used to justify the nation's political and economic strategy for becoming an important actor in the global system, the South African red meat chain shows how the same forces are used to build a modern (postcolonial) nation-state.[2] Unlike New Zealand, the export network in South Africa remains small, but it is steadily increasing. For export producers, who tend to be large and moving toward vertical integration, the traceability of the product from the farm to the dinner table is essential in gaining and maintaining an export certificate. Prior to 2000, the only HACCP systems in place were in the three European Union (EU) certified export abattoirs. Currently, the United States does not allow any raw red meat from South Africa to be imported because South Africa's food safety system does not satisfy the U.S. requirements for imported meat. Thus, some of the larger participants in South Africa are in the process of voluntarily implementing HACCP in order to further expand their export opportunities, both in Europe and in the United States.

At the other end of the spectrum lies the informal network common in rural South Africa. Much as in New Zealand, rural South Africans rely to a large extent on home-kill meat and informal slaughtering (i.e., slaughtering, cooking, selling, and consuming on site, usually under a tree or in the open air). But the majority of producers and consumers in South Africa fall within the domestic commercial network. After apartheid was ended in South Africa, economic liberalization was pursued as the new economic model. This meant significantly less government regulation of the agricultural sector. Thus, as the red meat chain shifted away from government control, the number of individuals participating in the red meat commodity chain increased, while there was simultaneously an increase in concentration and vertical integration in the export and higher-tier domestic commercial sector. Today, approximately 70 percent of red meat sold in the domestic commercial sector is supplied by feedlots, a 50 percent increase since the early 1990s. Similarly, starting in the late 1990s, as the nation gained membership in key international trade groups like the WTO, South Africa has begun to move slowly toward the HACCP-based food governance system using logic similar to that used in New Zealand. In 1999 the South African National

---

2. The case study of South Africa was conducted between September 2000 and August 2001. Approximately ninety-eight interviews with actors in the red meat commodity chain of South Africa were conducted, and documents related to the industry were collected and analyzed.

Department of Agriculture (NDA) introduced the Hygiene Assessment System (HAS) as "a standardized assessment method of hygiene standards in abattoirs." Modeled on the British HAS, this system gives abattoirs a score based on the level of hygiene in postmortem meat handling and penalizes those who receive substandard scores. This was an effort to move toward implementation of food safety standards at abattoirs while avoiding the high costs of HACCP and attempting to offset the increased consolidation and vertical integration occurring in the industry. (As costs rise, small operators are more likely to go out of business, which allows large operators to capture more of the market.)

In 2000 the South African government passed the Meat Safety Act, which stipulates the use of an independent inspection service. In March of that year, three industry organizations, the South African Meat Industry Company (SAMIC), the Red Meat Abattoir Association, and the National Emergency Red Meat Producers' Organization, jointly established the International Meat Quality Assurance Services (IMQAS) with the aim of becoming a designated inspection service provider under the Meat Safety Act. In June 2003 the National Department of Health established new regulations under the Foodstuffs, Cosmetics and Disinfectants Act of 1972 that require food-handling enterprises (or food handlers) to develop and operate with an HACCP system (SANDH 2003). In the future, the South African government hopes to follow the New Zealand model and establish a single food-control authority to administer various laws and regulations concerning food safety.

The South African case is significantly different from the New Zealand because the negotiations for food governance reform tend to be carried out among actors in the red meat chain who understand that South Africa is a developing country. The South African government has made food security for all South African citizens a top priority of domestic agrifood policy, and most actors within the commercial industry are aware of this priority. South Africa is currently not fully self-sufficient with respect to red meat (SANDA 2003). During an interview, one longtime researcher who works with the red meat industry said, "In the U.S., the consumer expects quality and meat is in excess, so they are unwilling to pay more, whereas here, meat is in shortage and therefore, if you want higher standards, then you will have to pay a higher price." The researcher's observation appears to correspond with the view of the majority of South African citizens who rank food access as their primary concern. Thus meat safety governance reform was driven by a desire to protect *citizens* and promote the growth of international *consumers*.

Tension between the national government and the meat industry persists, however. For export producers and large domestic producers the new food governance system is a necessary step for the future growth of the South African red meat industry. Yet the majority of actors who handle meat, whether from the domestic commercial or the domestic informal network, currently do not use a hygiene management system. This makes it extremely difficult for the NDA to implement the Meat Safety Act and other food safety laws. In part, the lack of hygienic meat-handling practices among actors in the domestic commercial network, who are overwhelmingly white, is due to increased costs, but it also is justified by racist stereotypes about their consumers, who are overwhelmingly black and poor. At the same time, industry groups such as SAMIC tend to blame the national government for hampering potential growth opportunities for the red meat industry. For example, the NDA has not yet approved IMQAS as a designated inspector.

What, then, is this reform for? It appears that there is no *agency* for South Africans in the reform of the food governance system, either as consumers or as citizens. Do South African citizen-consumers have any role to play?

Red meat chain actors, especially the larger producers and processors of the domestic commercial and export sector, attempt to enroll consumers to justify their actions. Take, for example, a campaign launched by the industry feedlot association, as described by an online marketing business Web site: "The South African Feedlot Association, this week launched a first-of-its-kind advertising campaign that will educate, persuade and assure consumers of the hygiene and wholesomeness of South African Beef as well as its health and nutritional value, appetite appeal, versatility and affordability" (Write Agency 2003). A description of the campaign can also be found at the feedlot industry association's Web site (www.safeedlot.co.za).

The red meat chain actors have attempted to determine how the three food networks intersect, and they defend their view about how to balance the nation's goals to achieve food security and harmonize food safety standards with those set by international organizations and overseas governments. As a meat inspector at SAMIC pointed out, "The government's first priority is food security [for South African *citizens*]. You cannot have food security with expensive food; cheap food must be available. And you cannot give [South African *consumers*] cheap food with high standards, because standards cost money. Ninety percent of the meat industry is in a Third World situation, while the other 10 percent cater to elite markets. It will be quantity with a low price that is the goal for that 90 percent. Ten percent will have high prices and be focused on export markets."

Yet the establishment of food safety regulations is necessary to protect the domestic red meat chain from increased overseas competition and to protect South African citizens from being marginalized in the global red meat system. A representative of the South African Bureau of Standards stated, "Quality is important, but is not the primary concern. Quality standards are in place primarily to keep [overseas] countries from dumping poor quality products on the South African market."

Policy and institutional reforms initiated by both the government and the red meat industry—such as measures to improve production and protect the domestic industry from overseas competition—have been overwhelmingly producer-oriented (domestic commercial). Yet consumers are seen as a key group in raising food safety standards. A representative of a meat import-export group stressed the importance of multinational supermarkets in standardizing consumers' expectations for meat quality and safety: "Changes are being brought about by bigger retailer groups like Pic 'n' Pay, Woolworth's, Spars, who use first world standards, and they are expanding. They are probably selling 65 percent of the meat. The other guys (small butchers, bush slaughtering, farm slaughtering) their standards have dropped, they have Third World standards." With a rapid increase in the number and market share of supermarkets in eastern and southern Africa (Weatherspoon and Reardon 2003), consciousness about food safety is increasing among white and black consumers alike. Actors who produce and circulate meat products in the domestic commercial network will be further differentiated by their ability to adopt and implement HAS, just as consumers are being differentiated by their ability to purchase meat products of varying quality.

While *citizens* continue to be concerned about access to red meat, the industry needs *consumers* to willingly embrace changes occurring in red meat production, distribution, and retailing. Many conventions surround the consumption of industrialized red meat, including the need for consumers to learn the names of specific cuts of meat and how different cuts of meat should be prepared. In addition, consumers must adjust to meat that tastes and looks different from meat produced in nonindustrial settings. Industrialization of the red meat chain also undermines traditional bases of trust (Lockie 2002). Standards and standardization (e.g., HACCP) are part of the increasing industrialization of the red meat chain, and consumers must accept the increasing distance between production and consumption practices and be willing to adjust their conception of *trust* from an individual to an institutional relationship. In rural areas, people rely on their trust in a local bush slaughterer that the meat he is selling is safe. As an industry adopts standards and standardization for the purpose, in part, of international trade, the industry must simultaneously persuade domestic consumers to trust institutions over and above individuals. Only by securing the participation and trust of the majority of consumers can the industry successfully implement consistent industry-wide standards. Finally, by winning the cooperation of South African consumers, the industry and the government further their ability to control animal diseases like measles and foot-and-mouth disease that, if left unchecked, inhibit South Africa's export potential.

In short, the reform of the food governance system in South Africa does not consist merely to allow a few large-scale producers to participate successfully in the global market. It must also secure the participation of previously marginalized consumers, while at the same time making the state define and protect citizenship rights to food. While *consumers* are increasingly differentiated on the basis of socioeconomic status, the *citizens* of South Africa will be further integrated by overcoming race as the fundamental category for redistribution of economic wealth, political power, and social status. In the postapartheid era, the state is expected to be responsible for improving land and capital distribution among black South African citizens who hope to become commercial producers. At the same time, the state is expected to support the development of the red meat chain through measures to promote meat export, while protecting the domestic chain from international competition and deadly animal diseases such as FMD and BSE and foodborne pathogens such as *E coli O157:H7* and salmonella. Despite the nation's status as a developing country, negotiations for the new food governance system take place within the red meat chain in the effort to answer the question whether South African citizen-consumers will have access to meat in general and safe meat in particular. Despite the significant level of socioeconomic inequality among South African consumers, they are far from powerless. Consumers' power is situated in the fact that the industry must enlist the cooperation of consumers, because without their cooperation the industry restructuring will not be successful.

Conclusion

Through comparative analysis of meat safety governance reform in New Zealand and South Africa, this chapter has attempted to show how agency is constantly reshaped and transformed, as structural features of the global agrifood system continue to change. On the one hand, the efforts of these countries to incorporate the HACCP system is a response to emerging structural features of the global agrifood system that has the potential to transform their predominant agrifood systems fundamentally. On the other hand, our case studies also show that this transformation is not necessarily uniform across time and space. This is because institutional methods of bringing food from farm to dinner table are built through historically contingent processes in a given place over a given time. The red meat chains in New Zealand and South Africa are the product not only of geographical location and ecological conditions, but also of sociocultural and political histories of negotiating how to produce, distribute, regulate, and consume red meat in these countries.

As many commodity chain studies (e.g., Bonanno and Constance 1996; Busch et al. 1991; Dixon 1999; Friedland 1984, 2001; Heffernan 1984) show, commodity chains are a vehicle for organizing the state and the market. But the relationships that consumers have with the state and the market are different from those that citizens have, though there is a considerable overlap. In everyday practices of producing, distributing, regulating, and consuming beef or lamb, individuals and groups are situated in a web of social relations and become social actors with multiple positions and roles. By differentiating between *consumers* and *citizens,* we can analyze how social actors strategically differentiate their roles to express their agency to shape their relationships to the market and the state.

In our case countries, meat safety governance reform aims to transform the roles of both state and market in regulating meat safety, and therefore the relationships that consumers and citizens have with these social institutions. As Marsden, Flynn, and Harrison (2000) suggest, this new food governance system *delegates* the state's traditional responsibilities to private companies—as seen, for example, in the development, implementation, and maintenance of HACCP plans. This appears to give meat companies greater accountability in maintaining the safety of food supplies in order to protect their consumers, while allowing them more flexibility in responding to the diverse food safety requirements imposed by governments and companies overseas. But this market self-governance approach also indicates that regulating meat safety is no longer about a *state's* responsibility for protecting *citizens'* rights to safe food, but instead about a *market's* responsibility for protecting *consumers'* rights to safe food.

Equally important is the state's role in fostering commerce in the increasingly global economic system. To maintain social stability, it is vital for the state to invest resources for the enhancement of economic prosperity for its citizens. In the process of negotiating reform, asymmetries among the three food networks mentioned above are exacerbated. Participants in the red meat commodity chain in New Zealand and South Africa do not benefit equally from reforms. New regulations threaten the existence of informal producers and processors and potentially impair the livelihoods of these citizens. Such asymmetries are nothing new. Numerous agrifood studies have repeatedly suggested that unequal distribution of economic resources and political power within a given commodity chain, or within the agricultural sector as a whole, is a structural feature of capitalist economies. But it is important to stress that the benefits and costs of a given reform are distributed very differently among countries and over historical periods.

In the case of incorporating the HACCP system, the central question is how to ensure the safety of meat. In order to answer that question, each actor in the red meat chain must also determine who are *the consumers* to be protected and who are *the producers* who must protect consumers from unsafe meat. In our case countries, consumers are differentiated into informal, commercial domestic, and three export market segments. In New Zealand, the differentiation is most notable between overseas and domestic consumers. In South Africa, the differentiation between rich and poor consumers is most apparent. Despite structural changes beyond their control, consumers and citizens are far from powerless. While consumers may exercise their agency through purchase choices in impersonalized markets, the reform of food safety governance in New Zealand and South Africa provide citizens with an opportunity to bring together their individual voices as consumers to express a *collective* form of agency through the political process that demands of the state the right to safe food. While consumers may be differentiated by their consumption patterns, citizen participation in public policymaking is necessary in order for the structural changes to succeed.

Agrifood studies should not take categories like "producers" and "consumers" for granted, nor should they ignore the importance of citizens as a viable category in the increasingly global agrifood system. Empirical studies must clarify how certain producers and consumers become participants in global agrifood markets while others are excluded from them; and how citizens in a particular nation participate in restructuring global agrifood markets. Only then will we be able to understand how tensions between structure and agency in the increasingly global agrifood system are articulated in a given place and time.

REFERENCES

Bonanno, Alessandro, Lawrence Busch, William H. Friedland, Lourdes Gouveia, and Enzo Mingione, eds. 1994. *From Columbus to ConAgra: The Globalization of Agriculture and Food*. Lawrence: University Press of Kansas.

Bonanno, Alessandro, and Douglas H. Constance. 1996. *Caught in the Net: The Global Tuna Industry, Environmentalism, and the State*. Lawrence: University Press of Kansas.

Bristow, Robyn. 1999. "New Law Heaps Major Costs on Oxford Abbatoir." *The Press* (Christchurch, New Zealand), 6 October, 4.

Brom, Frans W. A. 2000. "Food, Consumer Concerns, and Trust: Food Ethics for a Globalizing Market." *Journal of Agricultural and Environmental Ethics* 12 (2): 127–39.

Busch, Lawrence, William Lacy, Laura Lacy, and Jeff Burkhardt. 1991. *Plant, Power and Profits: Social, Economic, and Ethical Consequences of the New Biotechnologies*. Oxford: Basil Blackwell.

Cutt, John. 1998. "Replacement of 1981 Meat Act Welcomed." *Southland Times* (Invercargille, New Zealand), 23 December, 21.

Dixon, Jane. 1999. "A Cultural Economy Model for Studying Food Systems." *Agriculture and Human Values* 16 (2): 151–60.

Draper, Alizon, and Judith Green. 2002. "Food Safety and Consumers: Constructions of Choice and Risk." *Social Policy and Administration* 36 (6): 610–25.

Dyck, John H., and Kenneth E. Nelson. 2003. *Structure of the Global Markets for Meat*. U.S. Department of Agriculture, Economic Research Service. Washington, D.C.: U.S. Government Printing Office.

Food and Agriculture Organization of the United Nations (FAO). 2003. *FAOSTAT Agriculture Data*. 28 May. http://faostat.fao.org/.

Foucault, Michel. 1979. *Discipline and Punish: The Birth of the Prison*. New York: Vintage Books.

Friedland, William H. 1984. "Commodity Systems Analysis: An Approach to the Sociology of Agriculture." In *Research in Rural Sociology and Development*, vol. 1, ed. Harry Schwarzweller, 221–35. Greenwich, Conn.: JAI Press.

————. 2001. "Reprise on Commodity Systems Methodology." *International Journal of Sociology of Agriculture and Food* 9 (1): 82–103.

Gabriel, Yiannis, and Tim Lang. 1995. *The Unmanageable Consumer: Contemporary Consumption and Its Fragmentations*. London: Sage Publications.

Goodman, David. 2001. "Ontology Matters: The Relational Materiality of Nature and Agro-Food Studies." *Sociologia Ruralis* 41 (2): 182–200.

Goodman, David, and Michael J. Watts, eds. 1997. *Globalising Food: Agrarian Questions and Global Restructuring*. London: Routledge.

Green Party of Aotearoa, New Zealand. 2003. "Greens in Time and Space: The History of the Green Party." www.greens.org.nz/about/history.htm.

Haan, Cornelis de, Tjaart Schillhorn van Veen, Brian Brandenburg, Jerome Gauthier, Francois Le Gall, Robin Mearns, and Michel Simeon. 2001. *Livestock Development: Implications for Rural Poverty, the Environment, and Global Food Security*. Washington, D.C.: World Bank.

Heffernan, William D. 1984. "Constraints in the U.S. Poultry Industry." In *Research in Rural Sociology and Development*, vol. 1, ed. Harry Schwarzweller, 237–60. Greenwich, Conn.: JAI Press.

Juska, Arunas, Lourdes Gouveia, Jacki Gabriel, and Susan Koneck. 2000. "Negotiating Bacteriological Meat Contamination Standards in the U.S.: The Case of *E. coli* O157:H7." *Sociologia Ruralis* 40 (2): 249–71.

Krippner, Greta R. 2001. "The Elusive Market: Embeddedness and the Paradigm of Economic Sociology." *Theory and Society* 30 (6): 775–810.

Le Heron, Richard, and Michael Roche. 1999. "Rapid Reregulation, Agricultural Restructuring, and the Reimaging of Agriculture in New Zealand." *Rural Sociology* 64 (2): 203–18.

Lockie, Stewart. 2002. "The Invisible Mouth: Mobilizing the Consumer in Food Production-Consumption Networks." *Sociologia Ruralis* 42 (4): 278–94.

Long, Norman, and Magdalena Villarreal. 1998. "Small Product, Big Issues: Value Contestations and Cultural Identities in Cross-border Commodity Networks." *Development and Change* 29 (4): 725–50.

Marsden, Terry, Andrew Flynn, and Michelle Harrison. 2000. *Consuming Interests: The Social Provision of Foods.* London: UCL Press.

McMichael, Philip. 1998. "Global Food Politics." *Monthly Review: Hungry for Profit—Agriculture, Food, and Ecology* 50 (3): 97–111.

Millstone, Eric, and Patrick van Zwanenberg. 2002. "The Evolution of Food Safety Policy-Making Institutions in the UK, EU, and Codex Alimentarius." *Social Policy and Administration* 36 (6): 593–609.

New Zealand Food Safety Authority (NZFSA). 2002. "Strategic Direction." www.nzfsa.govt .nz/about-us/strategic-direction/index.htm.

New Zealand Meat. 2003a. "New Zealand Beef, 2003." www.nzmeat.co.nz/wdbctx/ beefus/beefus.home.

———. 2003b. "New Zealand Lamb, 2003." www.nzmeat.co.nz/wdbctx/lambus/lambus .home.

———. 2003c. "The Business of New Zealand Meat, February 2003." Wellington, New Zealand.

New Zealand Ministry of Agriculture and Forestry (NZMAF). 2003. "Exports." www.maf .govt.nz/statistics/internationaltrade/exports/index.htm.

Ollinger, Michael, Danna Moore, and Ram Chandran. 2004. *Meat and Poultry Plants' Food Safety Investments: Survey Findings.* U.S. Department of Agriculture, Economic Research Service. Washington, D.C.: U.S. Government Printing Office.

Public Broadcasting Service (PBS). 2002. "Modern Meat." *Frontline.* Aired 18 April 2002.

Ritzer, George. 1992. *The McDonaldization of Society: An Investigation into the Changing Character of Contemporary Social Life.* Newbury Park, Calif.: Pine Forge Press.

Roberts, Donna, and Laurian Unnevehr. 2003. "Resolving Trade Arising from Trends in Food Safety Regulation: The Role of the Multilateral Governance Framework." In *International Trade and Food Safety: Economic Theory and Case Studies,* ed. J. C. Buzby, 23–47. Washington, D.C.: U.S. Government Printing Office.

Royal Commission on Genetic Modification. 2001. *The Report.* Wellington, New Zealand: Royal Commission on Genetic Modification.

Scholosser, Eric. 2002. *Fast Food Nation: The Dark Side of the All-American Meal.* New York: HarperCollins.

South African National Department of Agriculture (SANDA). 2003. "Annual Report of the Director-General, National Department of Agriculture, 1999." www.gov.za/ annualreport/1999/agric99.pdf.

South African National Department of Health (SANDH). 2003. "Regulations Relating to the Application of the Hazard Analysis of Critical Control Point System (HACCP System)." *Government Gazette of the Republic of South Africa* 456 (27 June): 3–22. www.doh.gov.za/docs/regulations/index.html.

Southland Times (Invercargille, New Zealand). 2000. "Slaughter Deadline Arrives in Two Weeks." 15 April, 22.

Statistics New Zealand/Te Tari Tatu. 2003. *2001 Census Snapshot 10: Rural New Zealand.* www.stats.govt.nz/.

U.S. Department of Agriculture, Foreign Agricultural Service (USDA-FAS). 2003. "Meat: Beef and Veal." 20 March. www.fas.usda.gov/.

Walters, Mark Jerome. 2003. *Six Modern Plagues and How We Are Causing Them.* Washington, D.C.: Island Press.

Weatherspoon, David D., and Thomas Reardon. 2003. "The Rise of Supermarkets in Africa: Implications for Agrifood Systems and Rural Poor." *Development Policy Review* 21 (3): 333–55.

World Bank. 2005. "2.1 Population Dynamics." *World Development Indicators 2005.* www .worldbank.org/data/wdi2005/pdf/Table2-1.pdf.

Write Agency. 2003. "South Africa's Feedlot Industry Launch Campaign." 23 October. www.biz-community.com/Article.aspx?c=11&l=196&ai=2636.

INSTRUCTOR'S RESOURCES

Key Concepts and Terms:
1. Global agrifood system
2. Commodity chain
3. Food networks
4. Hazard analysis and critical control point
5. Constructed consumers, constructed producers

Discussion Questions:
1. Why do the categories of producers and consumers become elusive in the global agrifood system?
2. Increasing the export of agrifood products is considered an important strategy for economic development in a given nation. Compare and contrast how the strategies of New Zealand and South Africa have positively and negatively affected actors in these countries.
3. What alternatives currently exist for people to purchase food without using the grocery store?

Agriculture, Food, and Environment Videos:
1. "Modern Meat." *Frontline,* Public Broadcast Service, 2002 (60 minutes).

Agriculture, Food, and Environment on the Internet:
1. United States Department of Agriculture, Food Service Inspection Service: www .fsis.usda.gov/.
2. New Zealand Food Safety Authority: www.nzfsa.govt.nz/.
3. World Trade Organization: www.wto.org/.

Additional Readings:

1. Dyck, John H., and Kenneth E. Nelson. 2003. *Structure of the Global Markets for Meat.* U.S. Department of Agriculture, Economic Research Service. Washington, D.C.: U.S. Government Printing Office.

2. Marsden, Terry, Andrew Flynn, and Michelle Harrison. 2000. *Consuming Interests: The Social Provision of Foods.* London: UCL Press.

3. Scholosser, Eric. 2002. *Fast Food Nation: The Dark Side of the All-American Meal.* New York: HarperCollins.

CONCLUSION: FROM MINDFUL EATING
TO STRUCTURAL CHANGE

Wynne Wright and Gerad Middendorf

We began this book by laying out some of the major changes we see taking place in our relationship to food and agriculture. From the contentious politics aimed at rectifying injustices around global trade to the debates over scale-appropriate production and processing, contests over the future of food and agriculture are thriving. Few topics evoke the passion and controversy as do those involving our food. Because of the breadth and intensity of these contests, we contend that substantive change is taking place in our food system. If we gage change by the degree of communication, conflict, and tension surrounding a topic, then we indeed stand at the threshold of a new day in agriculture. Food is forcing us to raise socially uncomfortable questions about who we are, how we treat one another, how we coexist with nature, and who we will become as a society. The chapters in this volume speak to the diversity and depth of these changes. Perhaps more important, these chapters speak to the ways in which agency is exercised in contemporary society.

Contemporary agrifood systems entail new relationships and dynamic patterns of interaction that set us qualitatively apart from previous forms of agriculture. Our job as sociologists is to glean the patterns, threads, and commonalities that link, or segregate, these interactions. Given the changes described in this volume, we might expect to find unidimensional change. In other words, we might expect to find historical continuity in the agrifood system that achieved hegemonic status in the latter half of the twentieth century. On the contrary, the central theme we have identified is the emergence of a *contradictory agrifood system*. This contradiction emerges from the diversity of human agency.

We might conclude from the findings presented in these chapters that agency is being employed in such a way as to fashion a multitiered food system (Goodman 2004). On the one hand, we have seen global agribusiness giants continuing to pursue a bulk commodity food system based on the industrial manipulation of raw inputs. The fruits of the modernized productivist system are low-cost foods for the masses, with dysfunctional consequences hidden from the view of most consumers. The deleterious effects in this model are monumental, as the chapters in this volume have attested, but resistance to the industrial agrifood system should not be understated. Partly in response to the unsustainability of commodity agriculture, grassroots efforts aimed at regionalizing or embedding the agrifood system have become a growth industry. Some of these initiatives include agrifood systems that are imbued with multiple values and are more differentiated, less opaque, and more participatory than the dominant system. Finally, we note the existence of hybrids of these two forms of agriculture, such as Fair Trade networks. Early chapters in this volume described cases that drew our attention to the exercise of agency that appears to be bringing about some degree of systemic change.

The chapters by Johnston, Shreck, Munro and Schurman, and Skladany presented compelling examples of agency as a form of individual resistance, collective mobilization, and institutional realignment. Perhaps the most hopeful of all the chapters is Johnston's investigation of the Toronto-based good food box (GFB) scheme. Her work teaches us that agency directed toward reducing food insecurity can accomplish meaningful structural change, but can simultaneously be embedded with contradictions. At the same time that the GFB makes strides toward moving food away from its commodity form and reconstructing it as a basic human right, it can exemplify a form of "bourgeois piggery." Johnston found that the GFB contained elements of a counterhegemonic challenge to the global agrifood system, but at the same time it retained neoliberal principles of consumer sovereignty.

In Shreck's study of the Fair Trade banana network we see similar contradictory tendencies at work. While efforts to bring equity into the marketplace are growing, transformative change remains somewhat remote; little penetration into the status quo seems to be occurring. Like the participants in Johnston's GFB scheme, individuals were drawn to this consumption model as a way of bypassing the current system and redistributing wealth from the global North to the global South in the name of social equity. As it is currently configured, however, Fair Trade appears to present little challenge to the hegemony of the commodity system.

Munro and Schurman define agency as activism in the form of a collective and sustained critical analysis of agricultural biotechnology, which, upon consolidation, became the anti–genetic engineering (GE) movement. This social protest speaks volumes about the values that are driving changes in the food system and making political inroads toward challenging the industrial model. But the anti-GE movement is similarly riddled with contradictory tensions. Munro and Schurman caution against the expectation of significant social change in our food system via social activism that opposes biotechnology. By way of historical analysis, they contend that the anti-GE movement has been key in foregrounding concerns surrounding agricultural biotechnology. It has broadened its political and organizational network and drawn new actors into the fray. Regardless of these successes, the authors maintain that the movement is limited in its transformative range, given its failure to force "agricultural biotech into full-scale retreat."

Skladany's chapter highlights similar individual and activist approaches to the exercise of agency in the global salmon industry. He argues that in the case of salmon fisheries, contradictions emerge in the form of unintended consequences. Growing demand for salmon has presented consumers with a choice between industrially farmed salmon and wild salmon procured through more traditional methods. Choosing farmed salmon increasingly means participating in a production system with questionable consequences for human health and the environment, as well for indigenous cultures. Opting for salmon marketed as "wild," however, places responsibility squarely on the shoulders of consumers for participating in threatening the livelihoods of indigenous fishers and reducing the diversity of wild salmon stocks. What is one to do? These contradictory tensions suggest that one actor's agency is another's structural obstacle, according to Skladany's reading of the resistance and negotiation in the seafood arena.

Such contradictions are not confined to cases where agency appears to be making inroads in transforming the agrifood system. They persist in the final three case studies in this volume, where the common thread is the discovery of relatively minor departures from the ordinary or everyday owing to the structural impediments to transformative change. This is due, in part, to socioeconomic and political impediments to actor agency. In these cases, structure and culture restricted actors in their pursuit of food system change, reminding us that it is impossible to disentangle agency from structure.

Guptill finds that despite the combination of favorable ecological conditions, modern infrastructure, and committed, focused activists, organic farming in Puerto Rico remains negligible overall. Some gains have been made in connecting producers, consumers, academics, and policymakers around the promotion of organics and related alternatives, yet these efforts have not succeeded thus far in expanding the scope of a food system that holds the potential to challenge mainstream agriculture. Guptill's work shows that the structure of the Puerto Rican food system—historically dependent on food imports and nutritional assistance programs—reveals contradictory tendencies. It both enables and constrains the agency of activists seeking to transform the agrifood landscape. The Puerto Rican case, like others in this volume (e.g., Jussaume and Kondoh), highlights the importance of history and place in shaping the possibilities for activists to effect lasting, transformative change.

Jussaume and Kondoh examine food system change in four Washington State counties that have unique ecological, historical, and social contexts. They focus on "identifying the factors that shape human behavior within specific contexts of place and time" and "investigating the evolution of actors' attitudes and behaviors within these situations." They, like Guptill, find that structural possibilities for change vary from place to place, as do attitudes and behaviors. In Washington State it is this variability in context that creates contradictions that influence the possibility of reshaping agrifood systems.

In deconstructing the concepts of consumer and citizen, Tanaka and Ransom find that contradictions emerge in the form of structural obstacles to the proliferation of individual agency, while at the same time new openings emerge for the exercise of collective agency in the case of New Zealand and Africa. As domestic consumers in each of these markets were shut out of global red meat supply chains, opportunities emerged for citizens to engage in collective organizing and the making of public policy. By differentiating consumers and citizens, they are able to see that the multiple—and often competing—roles actors play in social life is fundamental to understanding the variable ways in which agency may be expressed, or repressed. By the same token, these contradictory roles of consumer and citizen alert us to the responsibility of considering the contradictory tensions between the multiple roles actors may play in the development of an agriculture system whose hallmark is biological, social, economic, and cultural diversity.

At this point we have to ask ourselves, "Why all the contradictions?" The authors of this volume are not alone in identifying tension and paradox in the food system. Others have discovered similar "dueling" principles, or contradictions, embedded in agricultural restructuring (Allen 2004; Goodman 2004; Hinrichs 2003; Holloway and Kneafsey 2000; Mooney 2004; Morris and Evans 1999; Winter 2003; Wright 2005). These chapters have underscored the complexity of our relationship to food and agriculture by providing empirical evidence of the ways in which conventional social and spatial patterns shape and inform the emergence of a "new" agriculture. We see this body of work contributing to Goodman's (2004) call for sociologists to probe the "ways in which the 'old' might reshape the 'new.'" The tension and contradiction demonstrated in each of these cases suggest that the claims about the eclipse of modernization on the farm and in our food system may be premature, as witnessed by the highly contested nature of change and the persistence of modern production practices alongside emergent forms.

No doubt, a host of variables are responsible for these contradictions, including power imbalances among various actors in the food system, the social reproduction of historical patterns, flawed public policy, overreliance on technology as a problem-solving agent, and undeveloped (or flawed) knowledge systems, to name a few. Our goal in these last pages is to draw attention to the sociocultural and organizational dynamics that stymie individual and collective agency.

Many of these contradictions are a product of socially disembodied eaters. By this we mean that these initiatives are struggling to achieve meaningful change, in part, because they collide with a culture that mitigates the multiple roles of individuals and privileges economic dimensions over other facets of social life. For example, we frequently eat from our social class location (Bourdieu 1984). We consume what we can afford. We sometimes stretch the boundaries of the economic dimensions of social class and consume things outside our economic reach in order to construct an identity that we want to project to others. Or we consume things out of custom, things that we like, that taste good, that make us feel good, and that are sold to us within a range that we can manage financially. All of these motivators remind us of the complexity of market exchanges that cannot be reduced to economic rationalization.

But, historically, we have artificially separated our consumer action from our citizenship role, as Tanaka and Ransom make clear. Few of us make consumption decisions while factoring in our values about the physical world and the impact that our economic agency has on our environment. Rarely are we conscious of ourselves as workers, so the labor equity issues of others get pushed to the back burner. Race, gender, and other inequalities on which the current agrifood system is built get obscured in this fragmentation of social life (Allen 2004). In short, we consume as undersocialized agents neglecting the social relations in which we are embedded (Granovetter 1985).

If meaningful agency is to be exercised—agency that can fundamentally shake the roots of the agrifood system and move us toward more sustainable development—these chapters suggest a need to reintegrate the multiple dimensions of our social selves over the rational calculation typified by the conventional agrifood system, which reduces us to economic actors and strips us of other social roles. Compartmentalizing social roles in this fashion has led to a hierarchical arrangement that privileges economic action at the expense of citizenship. The utilitarian logic of agricultural markets must give way to a more embedded perspective that encourages a thoroughgoing analysis of social relations, structures, historical forces, and situational constraints (Granovetter 1985). A more socialized view of agrifood systems would illuminate the nesting of individuals in society and free us to exercise our noneconomic roles. A similar perspective is laid out by the communitarians (Bell 1995; Etzioni 1996), who place primacy on the health of civic life, which thrives when we understand ourselves as social beings bound up in a multitude of networks.

Once we view consumption and production as inherently social, with multiple dimensions and unintended consequences, we can begin to uncover the assumptions embedded in the current—and future—food system and become *mindful eaters.*[1] Mindful eaters are reflexive eaters. They posses a quality of engagement with their food system that allows them to draw the linkages between the vast number of actors in the food system who span both production and consumption spheres. Mindful eating facilitates the unveiling of taken-for-granted assumptions and fosters the development of individual, collective, and institutional agency directed toward the twin goals of breaking down the atomized consumer of neoclassical economics and nurturing the socially engaged citizen. It reveals how meanings around food and agriculture are constructed, how agrifood in-groups and out-groups are distinguished, how farm labor is rewarded, and how race and gender inequality is reproduced, among other things. Finally, mindful eating allows us to locate our decision making within a broader social context and to act on our multiple roles. Food-related actions, then, become merely one kind of decision among the many we make within a broader social context. Mindful eating can bring an end to consuming from our class locations and facilitate agency that can balance both economic and noneconomic values. In short, we have to eat our values, but our values have to be driven by our commitment to public life and civic engagement.

Many are well on their way to individual change by bringing their values to bear on their food choices. Individual agency exercised via mindful eating may move us out of the starting gate, but it cannot go the distance. Mindful

---

1. This concept is built upon the work of Michael Schwalbe's (2004) notion of sociological mindfulness.

eating helps construct the social and cultural context in which individual agency can flourish, and it begins to set the stage for systemic change. But refashioning, or dismantling, historically embedded social structures requires collective agency aimed at institutional change. Even widespread consensus on the importance of eating mindfully would not magically revolutionize the food system. Even in a highly consensual context, the lack of a social infrastructure and institutional resources would inhibit social change, as would the tendency for some to free ride in a climate of widespread enthusiasm (Schwartz and Paul 1992).

Food and agriculture have to move beyond the elite Tuscan-inspired kitchens graced by celebrity chefs like Rachel Ray and Emeril Lagasse and their legions of devoted followers. Food and agriculture has to leap from the pages of the *New York Times* and the domain of high-profile writers like Michael Pollard and Barbara Kingsolver. They have to penetrate our national dialogue and gain a prominent place in cultural life. A "new agriculture" must fundamentally shift away from the market toward a balance between market and society. It must serve multiple social needs and allow participants to act upon multiple social roles. Lyson's (2004) work begins to get to the heart of these objectives. Lyson proposes a "civic agriculture" that embeds social and economic relations in their production context. Civic agriculture exists between market and society; it is a hybrid production form because it is both a mechanism for making a profit and a community-building tool. It serves multiple social needs. It allows individuals to participate in the market as consumers, but also as citizens concerned about the environment, health, labor, and other social issues. Some have argued for moving food and agriculture out of the economic sphere altogether and viewing food as a basic human right—not another commodity to be bought low and sold high. Civic agriculture retains the need for profit yet integrates other social needs such as solidarity, integration, literacy, equality, and democracy, thus blurring economic and social boundaries. This approach departs from the model of the economically rational actor and recognizes the multiplicity of roles individuals play as producers, eaters, voters, neighbors, parents, environmentalists, churchgoers, and so on, thus allowing agency to be exercised in a broader socioeconomic context.

How do we bring this civic food system to the table? We can pull up a chair gently, as those who advocate incremental change and behavior modification prefer. Or we can overturn the table and pursue the strategies of oppositional politics. We leave it to others to present a more thorough analysis of the merits of each of these approaches—incremental adaptation versus contentious politics. Without a doubt, both are fruitful lines of inquiry for future research. For our purposes, we conclude with one primary lesson.

Clearly, everyone has to have a seat at the table, regardless of what shape the table takes. Any effort to catalyze change must be a collaborative endeavor that weaves together the hungry, eaters, farmers, environmentalists, policy-makers, processors, distributors, educators, entrepreneurs, and others. Whether the change taking place in the food system is an outgrowth of concern over national food security, pesticides in our drinking water, or the defense of small family farming and withering rural communities, these changes col-lectively illustrate the diversity of the food system. Vital assets for remaking our food system are located in the diversity of actors and their bold crea-tivity, powerful networks, and vital resources. The hidden function of this robust asset, however, comes in the conflict and contradictions that neces-sarily accompany such diversity. We err when we define agrifood conflict as a deficit to be corrected. Conflict is essential to our social well-being. It "is an expression of the very multidimensionality of things, the plurality of dif-ferent groups, interests, and perspectives that make up the world" (Collins 1994, 89). These things are an outgrowth of oppositions that "are not mutu-ally exclusive but in fact form a dynamic whole" (Mooney 2004, 80).

Following the Marxian tradition, Mooney argues for an embrace of insti-tutional friction or tensions arising from antagonistic opposing social forces as an elixir for sustainable development. He contends that, absent this "tension-filled balance of both political power and economic interests" (76), what re-mains is simply oppositional politics, which reduces society to a tit-for-tat game populated by winners and losers and is incapable of achieving sub-stantive change or sustainable development. "Given the necessity and ubiq-uity of such struggle, it is important that sustainable development be built upon institutions that can also sustain forms of struggle coincident with the value premises of our cultural heritage" (77). Social institutions must avoid the pitfalls of fragmentation and integrate space for struggle and paradox among multidimensional actors. It is only in the contradictions that the dynamism for meaningful change around agrifood system sustainability can germinate. Organizations that make institutional room for tension and debate can serve as an incubator for change (Schwartz and Paul 1992). To this end, we see the evidence of agrifood contests and paradox as a benchmark of promise.

We need public institutions that are receptive to the contradictory values at play in the agrifood system, that have the expertise to bring together the various stakeholders in order to shape innovative agrifood initiatives, and that also have the capacity to evaluate those efforts. The land grant universities, for example, are particularly well situated to be at the leading edge of these efforts. They have a fundamental mission to engage their constituencies with research, teaching, and outreach targeted at improving the quality of life of their constituents. Moreover, their traditional mission was aimed at serving the public good (Gerber 1997) and even enhancing democracy through education and outreach to empower common citizens (Middendorf and Busch 1997). The land grant universities could reinvigorate their mission by engaging the public in dialogue and collaboration that is reciprocal and inclusive, drawing not only on their faculty, staff, and students but on a broad range of constituents. They need to reinvent themselves as more engaged institutions, focused on serving the broader public good (Kellogg Commission 1999). By doing so, they could greatly contribute to a food system that is more sustainable, equitable, and democratic (Fear et al. 2007).

Whether actors are poised to reinvigorate the land grant system, storm the meetings of the World Trade Organization, or join a cow-share cooperative to access raw milk, ample opportunity remains for the exercise of human agency, as the chapters in this volume attest. We do not have to become "coolly accustomed" to hunger (Dréze and Sen 1991), inequality, the divisive forces of globalization, or the emergence of a class-driven food system. The studies in this volume indicate that space exists on a number of fronts for mobilizing agency toward meaningful change.

REFERENCES

Allen, Patricia. 2004. *Together at the Table: Sustainability and Sustenance in the American Agri-food System*. University Park: Pennsylvania State University Press.

Bell, Daniel. 1995. *Communitarianism and Its Critics*. Oxford: Clarendon Press.

Bourdieu, Pierre. 1984. *Distinction: A Social Critique of the Judgment of Taste*. Cambridge: Harvard University Press.

Collins, Randall. 1994. *Four Sociological Traditions*. New York: Oxford University Press.

Drèze, Jean, and Amartya Sen. 1991. *Hunger and Public Action*. Oxford: Oxford University Press.

Etzioni, Amitai. 1996. *The New Golden Rule*. New York: Basic Books.

Fear, Frank A., Cheryl L. Rosaen, Richard J. Bawden, and Pennie G. Foster-Fishman. 2007. *Coming to Critical Engagement: An Autoethnographic Exploration*. Lanham, Md.: University Press of America.

Gerber, John M. 1997. "Rediscovering the Public Mission of the Land-Grant University Through Cooperative Extension." In *Visions of American Agriculture*, ed. William Lockeretz, 175–86. Ames: Iowa State University Press.

Goodman, David. 2004. "Rural Europe Redux? Reflections on Alternative Agro-Food Networks and Paradigm Change." *Sociologia Ruralis* 44 (1): 3–16.

Granovetter, Mark. 1985. "Economic Action and Social Structure: The Problem of Embeddedness." *American Journal of Sociology* 91 (3): 481–510.

Hinrichs, C. Clare. 2003. "The Practice and Politics of Food System Localization." *Journal of Rural Studies* 19 (1): 33–45.

Holloway, Lewis, and Kneafsey, Moya. 2000. "Reading the Space of the Farmers' Market: A Case Study from the United Kingdom." *Sociologia Ruralis* 40 (3): 285–99.

Kellogg Commission on the Future of State and Land Grant Universities. 1999. "Returning to Our Roots: The Engaged Institution." www.cpn.org/topics/youth/highered/pdfs/Land_Grant_Engaged_Institution.pdf.

Lyson, Thomas. 2004. *Civic Agriculture: Reconnecting Farm, Food, and Community.* Medford, Mass.: Tufts University Press.

Middendorf, Gerad, and Lawrence Busch. 1997. "Inquiry for the Public Good: Democratic Participation in Agricultural Research." *Agriculture and Human Values* 14 (1): 45–57.

Mooney, Patrick H. 2004. "Democratizing Rural Economy: Institutional Friction, Sustainable Struggle, and the Cooperative Movement." *Rural Sociology* 69 (1): 76–98.

Morris, Carol, and Nick J. Evans. 1999. "Research on the Geography of Agricultural Change: Redundant or Revitalized?" *Area* 31 (4): 349–58.

Schwalbe, Michael L. 2004. *The Sociologically Examined Life: Pieces of the Conversation.* Mountain View, Calif.: Mayfield.

Schwartz, Michael, and Shuva Paul. 1992. "Resource Mobilization Versus the Mobilization of People: Why Consensus Movements Cannot Be Instruments of Social Change." In *Frontiers in Social Movement Theory,* ed. Aldon D. Morris and Carol McClurg Mueller, 205–23. New Haven: Yale University Press.

Winter, Michael. 2003. "Embeddedness, the New Food Economy, and Defensive Localism." *Journal of Rural Studies* 19 (1): 23–32.

Wright, D. Wynne. 2005. "Fields of Cultural Contradictions: Lessons from the Tobacco Patch." *Agriculture and Human Values* 22 (4): 465–77.

LIST OF CONTRIBUTORS

ALESSANDRO BONANNO is Professor and Chair of the Sociology Department at Sam Houston State University. In recent years Bonanno has studied the implications of globalization for social relations and institutions. He has investigated globalization's impact on the state, democracy, and the emancipatory options of subordinate groups. Bonanno has written eleven books and more than ninety refereed publications, which have appeared in English and other major languages. He is also the current president (2004–8) of the International Rural Sociological Association.

DOUGLAS H. CONSTANCE is Associate Professor of Sociology at Sam Houston State University. His research interests focus on the socioeconomic impacts of the globalization of the agrifood system on rural communities and issues related to sustainable agriculture. He is co-author of *Caught in the Net: The Global Tuna Industry, Environmentalism, and the State* (University Press of Kansas, 1996) and has written several journal articles and book chapters on these topics.

WILLIAM H. FRIEDLAND is Professor Emeritus at the University of California, Santa Cruz. He has researched and written about the globalization of agriculture and about agricultural commodity systems such as iceberg lettuce, tomatoes, grapes, and wine. He is co-editor of *From Columbus to ConAgra: The Globalization of Agriculture and Food* (University Press of Kansas, 1994) and of *Towards a New Political Economy of Agriculture* (Westview Press, 1991). His best-known work is *Manufacturing Green Gold: Capital, Labor, and Technology in the Lettuce Industry* (Cambridge University Press, 1981), a seminal work in the area of agrifood studies.

AMY GUPTILL is Assistant Professor of Sociology at the State University of New York, Brockport. Her chapter is based on her dissertation on the alternative agriculture movement in Puerto Rico. She has also studied farming in New York State and the U.S. Virgin Islands, trends in food retailing in the United States and Puerto Rico, and cooperative agricultural marketing. She is currently investigating the growth of the organic dairy industry in upstate New York.

JOSÉE JOHNSTON is Assistant Professor of Sociology at the University of Toronto. Her interest in food brings together several research threads, including globalization, political ecology, consumerism, and social theory. Johnston's work has appeared in such journals as *Theory and Society,* the *American Journal of Sociology, Signs: A Journal of Women in Culture and Society,* and *Agriculture and Human Values.* She has co-edited a book entitled *Nature's Revenge: Reclaiming Sustainability in an Age of Ecological Exhaustion* (Broadview Press, 2006), and is currently working on a book on food culture.

RAYMOND A. JUSSAUME JR. is Associate Professor and Chair of the Department of Community and Rural Sociology at Washington State University. His research focuses on the social, economic, and political impacts of trade and investment in food and agriculture, with an emphasis on understanding the degree to which different forms of social organization contribute to human development at the local level. He is the author of *Causes and Consequences of Japanese Part-Time Farming* (Iowa State University Press, 1991). He also has authored or co-authored twenty-two journal articles and nine book chapters.

KAZUMI KONDOH is a Ph.D. candidate in the Department of Sociology at Washington State University. Her research interests include climate change, community building, and sustainable agrifood systems. She is currently conducting research on urban farmland preservation in the context of local and global climate change in Tokyo, Japan.

NORMAN LONG is Professor Emeritus of the Sociology of Development at Wageningen University, The Netherlands. His long-standing theoretical concern has been to develop an actor-oriented analysis of development processes. His work focuses empirically on the dynamics of planned intervention, global/local transformations, commoditization and social value, and transnational migrant communities. He is the co-author of *Anthropology, Development, and Modernities: Exploring Discourses, Counter-tendencies, and Violence* (Routledge, 2000), and author of *Development Sociology: Actor Perspectives* (Routledge, 2001).

GERAD MIDDENDORF is Associate Professor of Sociology at Kansas State University. His research interests are in the areas of rural and environmental studies, the sociology of agriculture and food, international development, and science and technology studies. His recent work has included a study of information needs of organic growers and retailers, and a study of agrarian landscape transition in eastern Kansas. He has also published a number of articles and chapters on the implications of agricultural biotechnologies and on agricultural science and technology policy. Middendorf is currently developing projects on the role of Latinos in agriculture in the Great Plains.

WILLIAM A. MUNRO is Associate Professor of Political Science and Director of the International Studies Program at Illinois Wesleyan University. His research interests include the politics of state formation and development in the global South as well as popular organization in the international food economy. Munro has conducted field research on state politics in Zimbabwe and is currently researching the politics of democratization in rural South Africa. He is the author of *The Moral Economy of the State: Conservation, Community Development, and State-making in Zimbabwe* (Ohio University Press, 1998).

ELIZABETH RANSOM is Assistant Professor in the Department of Sociology and Anthropology at the University of Richmond. Her research interests include globalization and development, social studies of science and technology, agriculture and food, and gender studies. Ransom has published chapters and articles on these subjects in such journals as *Teaching Sociology* and *Monthly Review.*

RACHEL A. SCHURMAN is Associate Professor of Sociology and Global Studies at the University of Minnesota. Her areas of interest include agrarian studies, environmental and natural resource sociology, transnational studies, and political sociology. Schurman is currently working on a book (with William Munro) on the effects of social activism on the trajectory of the biotechnology industry. She is co-editor of *Engineering Trouble: Biotechnology and Its Discontents* (University of California Press, 2003).

AIMEE SHRECK is Research Director for the California Faculty Association, a labor union. Her research interests include the Fair Trade movement, the relationship between sustainable agriculture and social justice, farm working conditions, and, more recently, labor relations in higher education. Shreck's chapter is based on dissertation research conducted in the Dominican Republic. Her research has been published in *Agriculture and Human Values, International Journal of Consumer Studies,* and other journals.

MICHAEL SKLADANY teaches sociology at the University of Tennessee, Knoxville. His research interests are in the intersections of science and technology with aquatic resource use, production, and consumption. His recent work has appeared in *Society and Natural Resources, Rural Sociology,* and *Monthly Review.*

KEIKO TANAKA is Associate Professor of Sociology in the Department of Community and Leadership Development at the University of Kentucky. Tanaka's research focuses on the role of science and technology in transforming social relations in agrifood systems. Her work has appeared in *Rural Sociology, Sociologia Ruralis, Southeast Asian Journal of Social Sciences,* and *Science, Technology, and Human Values.*

WYNNE WRIGHT is Assistant Professor of Community, Food and Agriculture at Michigan State University. Her research interests are in the contested nature of agrifood restructuring in global and local contexts. She is particularly interested in how people construct and negotiate meaning and exercise resistance to agricultural change. This work is shaped by public engagement with rural and agricultural communities. During the writing of this volume, she served as secretary of the International Sociological Association's Research Committee on Agriculture and Food (RC-40).

INDEX

accumulation, capitalist forms of, 50–51

Acott, Tim G., 227

active resistance: agency as form of, 56; passive resistance and, 70n.1

actor-network theory (ant), 51–53, 64–65; local agriculture revitalization and, 225–30, 243–44

actor-oriented theory: agency and, 81–83; anti-genetic engineering movement and, 147–72; Fair Trade networks and, 128–30; resistance and, 70–71, 76–79; salmon aquaculture and, 194–97

Adkin, Laurie, 98

affordance concept, social embeddedness of resistance and, 71–72

A/F Protein, Inc., 186–87

Africa, agricultural contributions from, 32

agency: agrifood systems and, 3–6, 45–66, 94–97, 273–81; in anti-genetic engineering movement, 146–72; case studies in, 18–20, 61; change created through, 18; classification, 46–47; of consumer-farmer alliances, 21; counterhegemony as expression of, 51–60; as defensive force, 8, 13, 54; evolution of, 47–48; Fair Trade movement and, 122–41; global agrifood systems and role of, 248–68; individual and collective modes of, 79–81; leverage issues in, 85–87; local agriculture revitalization efforts and, 225–30; Marxism and, 36–37, 48–51; meat safety governance and role of, 266–68; "mindful eaters" concept as tool of, 278–81; nature and, 45n.1; network effect of, 21; New Zealand meat safety governance and, 258–61; primitive agency, 51–53; in Puerto Rican food system, 204–22; research methodology concerning, 17–18; as resistance, 29–41, 62–63, 69–87, 274–81; salmon fisheries and aquaculture and, 177–78, 187–89, 193–97; socioeconomic conditions and, 36–37, 61–63; sociology of agriculture and, 14–17,

33–34; South African meat safety governance and absence of, 264–66; typology of, 60–63

Aggleton, Art, 105–6

agricultural biotechnology, challenges to, 13–14

Agricultural Development Board, funding regulations, 17

agricultural diversification funding, 16

agrifood systems: agency and, 3–6, 45–66; challenges to inevitability of, 11–12, 131; consumer-farmer alliances in, 21; contradictory characteristics of, 273–81; distancing from origins of, 101–2; divisions in Puerto Rico over, 216–19; Fair Trade movement vs., 124–41; globalization and transformation of, 69–87, 247–68; historical context for, 18; impact of anti-genetic engineering movement on, 148–49; local agriculture revitalization and, 227–30; meat safety governance and reordering of, 251–54; modern changes in, 38–40; in New Zealand, 255–61; organic production's impact on, 7–8; overiew of changes in, 2–6; professionalization of, 4; public awareness of, 3; in Puerto Rico, 8, 20–21, 203–22; reform efforts in, 20–21; resistance and agency, 29–41; sociological research on, 34–36; in South Africa, 261–66; structural change in, 83n.13, 273–81

Alamo-Gonzales, Carmen, 209

Alar pesticide crisis, as example of agency, 54–57, 63

Alaska, commercialization of salmon fisheries in, 180–81, 192

Allen, Edward, 145

Allen, Patricia, 122, 126, 276

Altered Harvest, 155

alternative food systems: commercial barriers in Puerto Rico to, 213–16; evolution of, 123–26; Fair Trade networks and, 121–41; liberatory and/or reactionary aspects of, 122; local agriculture revitalization and,

RURAL STUDIES SERIES

Clare Hinrichs, *General Editor*

The Estuary's Gift: An Atlantic Coast Cultural Biography
David Griffith

*Sociology in Government: The Galpin-Taylor Years in the
U.S. Department of Agriculture, 1919–1953*
Olaf F. Larson and Julie N. Zimmerman
Assisted by Edward O. Moe

Challenges for Rural America in the Twenty-First Century
Edited by David L. Brown and Louis Swanson

A Taste of the Country: A Collection of Calvin Beale's Writings
Peter A. Morrison

*Farming for Us All: Practical Agriculture and the
Cultivation of Sustainability*
Michael Mayerfeld Bell

*Together at the Table: Sustainability and Sustenance
in the American Agrifood System*
Patricia Allen

Country Boys: Masculinity and Rural Life
Edited by Hugh Campbell, Michael Mayerfeld Bell,
and Margaret Finney

*Welfare Reform in Persistent Rural Poverty:
Dreams, Disenchantments, and Diversity*
Kathleen Pickering, Mark H. Harvey, Gene F. Summers,
and David Mushinski

*Daughters of the Mountain: Women Coal Miners
in Central Appalachia*
Suzanne E. Tallichet

CPSIA information can be obtained
at www.ICGtesting.com
Printed in the USA
FFOW03n0311071215
19296FF